THE Translator of the following Treatise upon the Duties of Light Infantry in enclosed Countries, has thought that it may prove particularly useful at the present moment to the British Army. It will certainly be of great advantage to those who have not seen war, and it will recall to the memory of officers who have been upon actual service, many useful lessons.

It is to be understood by the most common capacity; and, to

be

be practised, requires only attention, activity, and zeal.

In such a country as England, the greatest utility will be derived from a numerous, active, and skilful light infantry, covering the positions and movements of the troops of the line, harassing the enemy's flanks and rear, attacking his detachments and convoys.

If the French attempt a landing in this country, they will, no doubt, endeavour to disembark a considerable body of troops of this description; and, indeed, all their troops are accustomed to fight *en tirailleur*.

Their

Their army will be constantly covered by sharp-shooters, concealed behind enclosures, hedges, trees, bushes, walls, houses, inequalities of the ground; they must be dislodged by a chain of English sharp-shooters, advancing under the same sort of cover, and driven behind their line.

The knowledge which we have of the country, must give us a great superiority over the enemy. The present Treatise contains lessons, and may create ideas which will enable the officers of light infantry to adapt their movements to the varieties of the ground

There

There cannot be the smallest doubt of the final discomfiture and total destruction of any army with which France has it in her power to invade Great-Britain: but we shall beat them with less trouble and bloodshed, in proportion to the perfection of our method.

L. A. M.

Late of the 8th or King's Inf.

INTRODUCTION.

LIGHT troops are in general intrusted with the out-duties of camps.

This kind of service is very extensive, whatever be the nature of the country; but its importance, with regard to light infantry, augments in proportion to the difficulty of employing cavalry.

In covered and inclosed countries, the surface of which offers a successive variety of heights and vallies; when the view is interrupted by woods, thickets, scattered trees, and a number of country houses;

houses; when the fields are surrounded by high hedges, and when you meet, besides, with ditches, ponds, morasses, &c. &c.: under such circumstances, almost the whole of the out-duties of camps must be performed by a numerous body of light infantry, and the long details of these duties seem to require a particular body of instruction, to enable light infantry to supply the deficiency of cavalry, where such local circumstances forbid almost entirely the use of that arm.

How necessary, then, for every light infantry officer to study attentively every part of the service, which concerns the safety of an army, either on a march or in camp, either with regard to foraging parties, escorting convoys, raising contributions in kind, and taking prisoners, or hostages, &c. &c.: especially with regard to reconnoitring the posts and positions occupied by the enemy, and giving notice of the marches or

move-

movements which he may undertake, either in totality or by detachments.

In war, great misfortunes have been occasioned by the want of sufficient instruction among subaltern commanders. It is, then, the duty of a man of honour, to qualify himself for discharging, with correctness and distinction, the duties of the situation he is in, and for acquitting himself of what he owes to his country. Whoever is invested with command in any capacity, ought not to look upon it as an honour unless he have made himself worthy of it: it is then that ambition is laudable; it is then that a noble emulation, guided and supported by military talents, becomes useful to the country, and that well-merited reward becomes an inappreciable honour.

The ignorance of military duties will lead into faults of which a single one may be sufficient to undo reputation, and

banish

banish for ever that confidence which
ought to take place between him who
commands, and those who are to obey,
and upon which alone depends success
and glory.

The duty of light troops being more
than any other performed by small detachments, in general commanded by
only one officer, instruction is the more
necessary to him, because every thing
that he wants he must find in himself,
and, in many perilous situations of that
so often insulated service, bravery, united
to military knowledge, can alone extri-
cate him with honour.

It is very requisite for an officer of
light troops, to obtain a thorough individual knowledge of the men under his
command, that he may employ them
according to their intelligence and courage.

One serjeant, corporal, or private,
will

will answer better for reconnoitring openly the enemy, that is, for approaching him, so as to be able to give a tolerable account of the post which he occupies and of his force.

Another will be better employed as a scout, or in watching the enemy's motions without discovering himself; and another will be found useful by his manner of questioning the peasants, and of getting from them the best information about the different regiments and uniforms, and the name and nature of the places where they lay, how far they send their patroles, their way of guarding themselves, the abundance or scarcity of provisions, forage, &c. &c. among them.

Another will be better calculated for an ambuscade, and have the necessary cunning for taking prisoners without compromising himself.

Others

Others are subject to infirmities, amongst which those of the sight must be particularly noticed; and even among them, some who see well in the day time are almost blind at night.

Some old soldiers have the genius of resources, and having observed some situation or passage, may be able to give a good advice, which ought to be turned to advantage.

And as some men are naturally awkward and easily alarmed, it is very important to know them, in order not to employ them where they might communicate their fears.

All those different characters may be easily found out by conversing with them, and chiefly by attending to their reports.

It is not only necessary for an officer of light troops, to obtain of those under him

him a perfect confidence in his judgment, his courage, and skill; but it is likewise important for him to gain their affections. He must know, that he cannot carry every thing by his single sword, and that if he is not heartily seconded in a service, where the readiness of the men influences, in a particular manner, the success, he never will be able to undertake or perform any thing without fear for his reputation.

However, the good will of the men is not to be obtained by the sacrifice of discipline, which is more particularly requisite with light troops, as they have more opportunities of escaping from its strictness.

The best manner of gaining the affections of the soldiers, without prejudice to discipline, is by providing carefully for their wants. It may sometimes be said, with regard to them, that necessity has no law; but that being over

its

its reign begins afresh, and in no case will individual plunder be of avail in procuring subsistence for the troops.

An officer must also distinguish when extreme vigilance is necessary, and when he can indulge his men with a little repose. Such attentions win the heart of a soldier, and he never will complain of an extraordinary exertion, when he is convinced it is necessary for his own preservation.

Drunkenness is a disgrace every where, but its effects are most dangerous at the out-posts. Nothing can ever wipe off the stain of having been surprised by negligence of common precautions. Temperance is a virtue, best inculcated by example: an officer who gets drunk himself, can scarcely punish a soldier for the same offence.

The greatest disgrace for an officer of light troops is to be thought greedy and

and eager for plunder : with such officers, no discipline can be maintained ; and the disgrace resulting from such a conduct, falls not only upon them, but attaches to the nation to which they belong.

Liberality and generosity are not military talents, but are virtues, without which military talents can never attain that high degree of glory and reputation, which leaves nothing to wish for.

The service of light troops, freed from whatever is repugnant to honour and morality*, without being useful to the

---

he was he

* The famous Pandour Trenk was not deficient in military talents, though his end affords a remarkable example.

It must besides be remarked, that in war every action cannot be regulated by the common standard of morality : the first object is success, and success often requires the sacrifice of what we owe to humanity.

A country is laid waste, the wretched inhabitants are unmercifully ruined, in order to prevent the enemy from being able to live in it, and of carrying the war farther. In such cases, the cattle, the corn, the

the ends of war, was always looked upon as the best school for officers, who are unwilling to die idle in their present rank, or in one a little higher, to which their ambition is restricted. The military career is boundless for great talents, and fortune often is prodigal of her favours to men of limited talents, who are strongly inspired with zeal and honour.

It is with light troops in particular that you will learn to examine carefully the inequalities of the ground; to make your dispositions accordingly, and to combine what can facilitate or oppose your

---

forage, are driven away, and the flames must destroy what could not be removed. But while you act in that manner with an enemy's country, you must not, in addition to it, ravish the married and unmarried women, nor torture the inhabitants, to force them to declare where they have concealed their money; nor deprive them of their cloaths, utensils, or furniture, which cannot be of any service to the enemy. In war, it is out of the question to distinguish theologically, what is just from what is unjust; but only what will promote or not the success of the war.

your undertakings. It is in this school that you will learn the necessary precautions for attack and defence; and it is here that you will be taught foresight, in the frequent retreats which light troops are obliged to make, from the variety of movements which take place.

A retreat performed before superior numbers, with such order, caution, coolness, and prudence, as circumstances allow, is one of the actions which characterizes most strongly the officer, and one of those which deserves most to be noticed by the generals, and the most proper to give them confidence to intrust him who performs it with more important enterprises.

It is in the midst of those multiplied events, that an intelligent and brave officer soon finds an opportunity of distinguishing himself. It is but seldom that a subaltern will find such an opportunity.

tanity. In battles and general engagements, the glory of those great actions must be divided among the generals; and, even for them, such opportunities are not frequent.

Having now shown, that the service of light troops is the quickest and surest road to preferment and honours, and indicated, in the mean-time, some of the most necessary rules, concerning the discipline and the moral character of a military man, we shall, in the following instructions, point out what conduct an officer is at all times to pursue in presence of the enemy, in order that he may deserve and obtain that praise due to ability, and which is so flattering to a soldier.

INSTRUCTIONS

SERVICE OF LIGHT INFANTRY

IN THE FIELD.

———o———

IT has been said in the Introduction, that light troops were, in general, entrusted with the out-duties of camps; and that, according to the nature of the country, infantry or cavalry had, respectively, a greater or less proportion of those duties to perform.

In treating particularly of the duties of light infantry, with regard to the exterior security of armies, and in distinguishing them from the duties of cavalry, it becomes necessary to establish the general connexion which exists between

B those

those two services; and to that effect, it will be proper to consider a camp in all its different parts, and the manner of guarding it, so that every one may have a complete idea of the duties which will be required of him, according to the nature of the ground, and other circumstances.

ON THE OUT-DUTIES OF CAMPS, AND THE GUARDING OF ARMIES.

Out-duties are of different kinds, and must be considered separately.

The first of these divisions will contain what concerns the chain of guards which extends round the camp.

SECTION I.

Of Guards intended for the Security of the Camp, and for the Prevention of Surprise.

When an army marches from one camp into another, it is always preceded by an advance-guard; its flanks are covered, according to circumstances, by a greater or smaller number of light troops, disposed of relatively to the situation of the enemy; and it is followed by a rear-guard.

When the advance-guard is arrived upon the ground where the army is to encamp, the flankers halt, and take such a position as is best calculated to cover the march, until the whole army be entered into camp with the baggage, and until the arrival of the rear-guard, which brings along with it all followers. Then the troops on the flanks gradually fall back and enter the camp,

B 2   along

along with the rear-guard; or rather, proceed to occupy the posts which have been pointed out to them.

The advance-guard is in general formed of all the troops, cavalry as well as infantry, which are intended to form the chain in front of, or on the flanks of the new camp.

This advance-guard is sometimes reinforced with a body of troops, which is to have a separate camp in front of the army; or for other purposes, relative to circumstances: but whatever be the force of the advance-guard, it is always followed by the camp colour-men.

The new guards and camp colour-men are under the command of the general officer of the day, whose duty it is to mark out the ground for the camp, and place the guards. That officer (after having given the line for the camp, marked the right and the left, and shown to the staff officers the head-quarters) is to march forward with the guards, the brigadiers and colonels of the day, as far

as

as he thinks proper ; to have the ground searched and reconnoitred, in order to satisfy himself that there is in the neighbourhood no concealed party of the enemy, either great or small.

After this first reconnoitring, he sends orders, as soon as possible, to the officers who have to mark out the camp, to begin their operations. While he advances and goes over the ground, he will examine what will be the extent of the chain ; he will point out to the brigadiers and colonels of the day who accompany him, the principal points of it, such as the heights, the woods, villages, and houses, where it is to pass, and the brooks and valleys that are to be left before it, and must be guarded. He will, likewise, point out to them how far the flanks of the chain are to be extended, and how they are to be supported; or whether, for want of a proper *appui*, it will not be necessary to encircle the camp, more or less, towards the rear.

That

That part of the chain which is intended to oppose the enemy at every avenue of the camp, ought to be distinguished from the guards placed in the rear of the army, and the object of which is to stop deserters and marauders. Such guards are placed after those which are to form the chain.

The extension and situation of the chain depend on several circumstances.

The first, concerns the distance from the enemy.

The second, depends upon the wants of the army. If the troops are to be supplied with forage, straw, wood, or even water, in front of the camp, the chain must be extended in proportion, and formed so as to cover and protect the troops, while they are engaged in providing for their wants.

A third consideration, although it results from the first, is that the most advanced guards ought not to be entirely
out

out of sight of the camp, that they may be more quickly supported in case of an attack. When the peculiarities of the ground prevent compliance with this rule, a small number of troops, infantry or cavalry, are incamped, or sent without tents to take post between the camp and the most exposed parts of the chain.

These different considerations are not always easily made to apply to the different natures of the ground: however it is the province of the major-general of the day, to mark the extent and situation of the chain, according to circumstances, and to place the first guards of cavalry*.

When the army is very considerable
. and

---

* To be more certain about the disposition of the chain, the major-general of the day ought to be assisted in this duty by some of the staff-officers, who could, in the mean time, sketch the ground with a pencil, remark the principal circumstances, and the number, force, and situation of the guards.

and the extent of the chain increases in proportion, the major-general of the day, to forward the work, entrusts the brigadiers and colonels of the day, who accompany him, with the detail of placing the different guards, dividing among them the chain, by right, centre, and left, or only by right and left, and taking care to fix the limits of each of these divisions by some conspicuous object.

It is the province of the brigadiers and colonels of cavalry to complete the reconnoitring of the country*, and to ascertain, more minutely, whether any party of the enemy be in ambush in front of the army, within the chain. They are also to place the advanced guards and posts of cavalry, upon the line which has been shown them.

The brigadiers and colonels of in-

fantry

---

* This is to be understood in open ground, where guards of cavalry may be employed.

fantry of the day, are to place the infantry destined to cover the camp, in such places as cannot be guarded by cavalry, and which, according to circumstances, may serve to cover the night-posts of cavalry. This is to be done in concert with the brigadiers and colonels of cavalry.

Every one having performed his task, he is to make his report to the major-general of the day, and to give him in writing the return of the guards, remarking, as well as the number of men, the roads, villages, woods, and heights, where the posts are placed.

The duty of the major-general is then to visit, himself, those different posts, and make such alterations in them as he thinks necessary.

Having made this inspection, he goes, himself, to report to the general in chief, and gives him a return in writing, of all the guards and posts, in order that he may divide the duty among

among all the different regiments * of the army, and that a copy may be given to the general officer of the day who succeeds him.

Such are the general dispositions concerning the chain; but some others, which depend on the nature of the ground, require more detail.

In marking the situation of the chain, it will be requisite to contain, in its circumference, as much as possible, any heights which might command a view of the camp, and offer to the enemy an opportunity

---

* When the army has a sufficient number of light troops to supply all the out-posts, the regiments encamped in line give only the camp guards, which are to be distinguished from the guards which serve to form the chain.

The camp-guards are only fifteen or eighteen men strong, and are placed about a hundred paces in front of the centre of every battalion of the first line, and at about the same distance in the rear of the centre of battalions of the second line. Such cavalry as is encamped has also camp-guards.

portunity of surveying its force, extent, and disposition, and what is going forward in it. If the chain could not, without inconvenience, be extended so far, it will be proper to examine whether a separate detachment could not be placed on that height, with proper precautions to support and insure its retreat if attacked.

It is disadvantageous to continue the chain through a wood. As much as possible the posts are to be placed on its exterior skirts, towards the enemy; and if this cannot be done, on account of the great extent of the wood, without requiring too many posts or weakening the chain, the guards must be placed, in that case, on some road which crosses the wood, or following the banks of a rivulet, or a valley, by which means they will have a better range to see before them.

Should the interior of the wood be very thick, and without rivulets or roads

in

in the direction of the chain, the best plan will then be to leave it before you, at a little more than musquet shot, and place in it, or upon any roads which may be in it, some small posts of infantry, or standing patroles, to give warning of the arrival of the enemy, as well in the day as in the night-time.

To carry the chain through a thick wood, where the sentries can neither see each other, nor stop spies or deserters, would be contrary to the first rules of this service. It is better, therefore, to leave it in front. If it happen, however, that the tail of the wood is too near the camp, the chain must be carried across it, and either an abattis must be made, or the posts of infantry increased in number.

The guards of cavalry must be placed, so as to prevent the enemy from being able to discover their principal post; and for that purpose it ought to be posted in a hollow way, or

on

or on the back of a rising ground, or behind a thicket, or house, about eight or nine hundred paces within the line of the vedettes.

Between the grand guard and the vedettes, a detached post is to be placed at about three hundred paces from them, but so that it may be distinctly seen from the grand guard, and see clearly all its own vedettes.

Every vedette must likewise be able to see the one on his right and left, and the intermediate ground between two vedettes must be seen at least by one of the two, so as to make it impossible that any thing should pass between them, without being perceived and challenged.

From the observation of this important rule results the perfect formation of the chain: and, to prevent spies or deserters from going through it, the vedettes must be double, so that one of the two may move and stop any one who might attempt to penetrate into the camp, or go out of it ; otherwise a single vedette,

vedette, who on no account is ever to quit his post, could do no more than call to the detached post, and the spy or deserter might easily get off before that guard could come up.

When, from local circumstances, vedettes must be placed at a greater distance than usual from the detached post, it is still necessary to have them double; for although the vedette might be easily seen from the guard, he might not be heard. In that case, one of the vedettes should go and acquaint the detached post with what has happened in front, which will then advance and reconnoitre; the vedette will immediately proceed to the grand guard with the same report.

In placing vedettes, it is to be observed, that they must not be too near the skirts of woods, or of any covered place which might be occupied by the enemy: they must be at the distance of six hundred paces from them.

It frequently happens, that from the dis-

disposition of the ground, one detached post is not sufficient to see all the vedettes which the grand guard furnishes to the chain, and to correspond with them: in that case, the vedettes must be divided, to correspond with a second detached post, which must be perceived from the grand guard, as well as the first.

The distribution of the guards must be made with great attention, to combine regularity in the performance of that duty with the smallest possible number of troops, as this is one of the most fatiguing duties for cavalry.

With regard to vedettes, they must be placed, in preference, upon heights, and on ground from which the view is extensive; at the entrance of roads, so as to command ravines, hollow ways, or other passages in front of the chain. If one or more trees should offer, from under which the country can be equally well discovered, they may be used to conceal the vedette from the enemy; but

but in order to see well, the advantage of concealment must very often be sacrificed.

With regard to the piquets of infantry, they ought always to be placed, so as to have protection from either natural or artificial fortifications.

The name of natural fortification is given to a church or church-yard, a garden surrounded by a wall breast high, or by a wet ditch, or any place that is not commanded within musquet shot, and surrounded with strong and impenetrable hedges, &c. &c.

Artificial fortifications are those that require some preparation, as cutting down trees, and making a circular abbatis, or piercing loop holes in a wall, raising a breast-work upon the inner side of a ravine or hollow way, or in short any intrenchment whatever with a ditch and breast-work.

The posts of infantry guards are the same for the day and the night: there must be one at the entrance of every

defile

defile and passage by which the cavalry might have occasion to retire.

The guards of cavalry and infantry ought to protect each other mutually; and should the infantry be in some places on the first line of the chain, the sentries of the right and left of such posts must be able to see and to be seen by the next vedettes.

Whenever the guards of cavalry retire during the night behind the posts of infantry, they must place vedettes on their flanks and rear for their own safety, and send out patroles to the front and flanks.

Should the ground be so open and bare of hedges and houses, as to be quite unfit for any post of infantry, the guards of cavalry will retire a little towards the camp when the night is very dark. The field-officers entrusted with the details relative to the chain, will point out to the guards of cavalry their day and night posts. Whenever the posts of cavalry on the chain happen to

be

be placed along a river, rivulet, or morass, the fords and nature of the bottom must be carefully reconnoitred.

. If the rivulet or morass be every where fordable, the guards of cavalry are to retire, as usual, to their allotted night-post.

But should it be positively ascertained, that the rivulet is only passable at certain fording places, and at the bridges, and the morass only by certain roads or foot paths, the vedettes must accordingly be planted opposite to these. In such cases, the guards continue in their post all night, and as soon as it becomes dark the vedettes move forward, quite close to such passage, &c. Should it be a wooden bridge, the planks must be taken off, but so as to be easily replaced if wanted.

When the banks of a rivulet are much covered, and its bed not fordable, guards of infantry should be placed in preference to cavalry, facing the passages, and should, of course, intrench them-

themselves in front, according to circumstances.

If the rivulet be found passable in too many places, and have an open plain behind, in that case it must be guarded by cavalry; but the vedettes must, in the day time, keep without musquet shot from its banks.

When the chain, through different motives, extends beyond that distance where the guards, in case of an attack, could easily be supported by the troops in camp, the most distant parts of it must then be supported by intermediate bodies of either infantry or cavalry, encamped in some situation calculated to answer that purpose.

In other circumstances, small posts may be pushed beyond the chain, to observe the motions of the enemy, and report them as quick as possible; but such irregular dispositions are not comprised among those that constitute the chain, all the parts of which ought to be connected, and form a cordon impenepenetrable to spies and deserters, and capable of baffling the reconnoitring of the enemy.

The duties which are treated of in this section are in general divided between the infantry and cavalry. In the following section, we shall treat separately of the particular conduct of light infantry officers, in the whole extent of the duties which they may be called upon to perform.

SEC-

SECTION II.

It has been already observed, that in certain countries the ground was so unfavourable to cavalry, that almost the whole of the out-duties of camps devolved upon the infantry.

The following instructions have been made upon this supposition, that from the nature of the ground, all the precautions relative to the guard and safety of armies fall to the lot of the light infantry.

On the Duties of an Officer commanding a Piquet of Infantry, or any other Out-Post.

The piquet having taken post at the appointed place on the chain, the commanding officer will place, in front and rear, the nearest and most obviously necessary sentries, whose position he may after-

afterwards alter, more or less, to agree with his general dispositions.

Meanwhile the piquet will remain under arms, facing towards the enemy.

On his return he will take with him about a third of his force, in order to reconnoitre exactly the whole of that part of the chain entrusted to his guard and care.

While he is reconnoitring, the remaining part of the piquet will continue under arms.

The officer commanding the piquet, either is or is not acquainted with the ground upon which he is placed.

If it be known to him, he will make his dispositions accordingly. At all events, it will be of service to him to have as good a map and spy glass as he can procure.

If he be not acquainted with the country, let it be his first care to get some of the inhabitants, from the neighbouring houses or village. With these guides he will proceed to the right of his ground, *viz.*
towards

towards the place from whence the first sentinel on his right ought to be able to see the first vedette or sentry on the left of the next post, so as to prevent any thing from passing between them, without being seen and stopped if found necessary.

While he is employed in reconnoitring and making his dispositions, he will send forward some small patroles of two or three men strong, who will regulate their march on his, keeping at the distance of two or three hundred paces before him, in order to prevent any thing from coming unawares upon him from the front or the flanks.

He will then, with his map in his hand, put questions to his guides upon the nearest objects represented on the map, and on all such as are within sight: he will carefully enquire the names of the hamlets and farms, woods, rivulets, heights, or considerable mountains, &c. If he can draw, he will take a sketch of the ground, marking the names of those objects,

objects, and the situation of his different posts; or he will write those names in his pocket-book, in the form of an itinerary, with their respective distances, so as to be able to answer, in a satisfactory manner, such questions as the generals may put to him, when they visit the out-posts; and, for his own sake, in case he should be attacked, he will inform himself, distinctly, from his guides, of every road in particular, and of every foot path within his post, and ascertain where they lead to; the nature of such roads from his post to the nearest villages: whether they are passable for carriages all the way, and consequently for cannon; whether they go through defiles, ravines, hollow ways, morasses, or woods; whether they are crossed by rivulets which are fordable, or are passed upon bridges, and whether those bridges are of stone or wood; whether those rivulets are fordable every where, or only at certain places.

Thus proceeding from the right to the left of his post, he will place his sentinels

tinels and small posts, in consequence of the information received.

We have already remarked, that all the vedettes or sentries on the first line of the chain ought to be double, and we have given the reason.

When the sentries can be distinctly seen from the principal post, and are not above three hundred paces distant from it, it will be sufficient to have them double, as we have said before; but should they not be clearly seen, or be at a greater distance than three hundred paces, it will be better to place on that spot a small detached post of four men and a corporal.

This small post will be placed behind a hedge, or on the skirt of a wood, so as to remain concealed, and will plant one sentry at about fifteen or twenty paces in front.

This sentry being so near the post, can easily acquaint the corporal with whatever he may see, who, after having reconnoitred himself, will send one of his

D

his men to the principal post, to inform it. This small post will relieve its sentry at the usual times, by day and night.

The commanding officer of the piquet having distributed his small posts and sentries, according to the information he has received from the guides, and to the nature of the ground, will be particularly attentive not to overlook any passage through which the enemy might come upon him; and in placing his posts and sentinels, he will point out to them the roads, ravines, woods, valleys, heights, houses, and villages, towards which they are to keep a constant look out. The sentinels of the out-posts, on the first line, ought to remain immovably turned towards that side which has been pointed out to them: they are not allowed to walk about in every direction, as those before a guard room. No light infantry soldier is ever, when on sentry at the out-posts, to support or slope his firelock, but to carry it advanced in his left arm,

as

as riflemen do. That position, called in German *gewehr beym fuss*, may also be used by sentries. The reason is, that in raising and moving their arms they are more likely to be seen: and every sentinel, placed for the purpose of observation, ought to endeavour, as much as possible, to see without being seen.

Another advantage resulting from the method of having double sentries is, the power of observing, from the same spot, different objects in different directions: thus, one of the two sentries will watch over one quarter of the horizon, and the other over the remaining part of its half circumference. The sight of a man does not diverge beyond ninety degrees, without turning the head.

The commanding officer having directed his posts and sentinels how to act, will give orders to the nimblest of the two sentinels to run to the principal post, and report any thing that may happen.

In the distribution of posts and sentries, the total number of men, thus employed

ployed

ployed to guard the whole extent of ground, must not exceed the third part or at the utmost the half of the strength of the piquet. If we reckon four men and a corporal to each small detached post, and two men for each sentry, the extent of the ground to be guarded might require three or four detached posts, and five or six sentries. Thus thirty-two men would be employed, only twelve of whom would be relieved from the principal post; consequently the piquet ought to consist of at least sixty rank and file, to supply and support, in case of emergency, that number of posts and sentries, upon an extent of fifteen or eighteen hundred paces of circumference, reckoning two or three hundred paces of interval between each post or sentry, that is, supposing the ground to be covered and crossed by several roads, the extent of an English mile upon the chain could not be well guarded with less than eighty men.

Having thus reconnoitred and distributed

tributed his posts, the commanding officer will return to his piquet: whereupon it is necessary to observe, that those different details may keep him employed for several hours. In that case, should every thing be quiet, and likely to continue so, the commanding officer will, as soon as he thinks it convenient, send word to that part of his piquet which he left at the principal post, that the men may lay aside their arms, but are not to be allowed to straggle.

The commanding officer having returned to the piquet, will, according to circumstances, and if he be not certain of the distance of the enemy, send out some patroles of four men and a corporal only. He will explain to them where and how far they ought or can go, and what precautions they are to take, in consequence of the information he received himself in reconnoitring.

The object of such patroles is in general to approach, undiscovered, the

villages

vilages or single houses on the road that has been pointed out to them, till an opportunity offer to speak to the country people, and question them upon what they know of the enemy :—where he keeps his nearest out-posts? How far he usually sends his patroles? Whether they know if he have made any movement? Whether they have seen any troops marching; in what numbers and in what direction? Should there be at any little distance from the road a height or hillock, likely to command a very extensive prospect, the non-commissioned officer commanding the patrole, may send there two of his men, such as have a good sight, to examine what is going forward in the vicinity, and as far as the sight can extend: mean-while he will conceal himself in some place shewn to them, until their return. If a party of the enemy approach towards them, they can give warning by whistling in a certain manner, or by making a signal with a pole, and then retire nimbly behind the

the hedges, and without going too far from the road. If the patrole acquire any important information, or reconnoitre any thing of consequence, the officer commanding the piquet will be sure to report it in writing to the field-officer commanding that part of the chain, and should a march and movement of the enemy have been distinctly ascertained, he will also send a report of it directly to head-quarters.

To forward intelligence with expedition, there ought to be with the piquets of infantry a certain number of light cavalry, for that purpose, and for other objects which shall be mentioned hereafter.

It has been remarked, that every post of infantry on the chain ought to be covered by some fortification, natural or artificial.

The first care of the officer commanding a piquet, when he arrives at his post, ought to be that of examining the ground, and seeing what could be
done

done to put it in a state of defence; but as he cannot fortify his post, and at the same time reconnoitre the ground, it will be sufficient for him, at first, to take some provisional steps to that effect, that is to say, that if he think it necessary to protect himself by an abbatis, he will mark out its situation and extent; he will leave directions to the person who is to command in his absence, to send to the neighbouring houses and villages for the necessary tools, and even the workmen. If it be proper to raise a parapet to stop a road, to pierce loop holes in a wall, &c. &c. he will give his orders accordingly. While he is reconnoitring the ground, the tools and workmen may be got, and, at his return, he can set them to work. Where workmen cannot be procured the soldiers must be employed.

To make himself equal to this important task, in every different situation where a piquet can be placed, an officer ought to have some knowledge of field-forti-

fortifiation; and should read, with attention, the small treatises which have been
written upon that part of the military
science.

It is, however, proper to remark
on this subject, that a piquet is not to be
considered as a post for resistance. A
piquet is essentially calculated to give
timely warning. The measures of defence for such a post are limited to the
necessary precautions to avoid a surprise,
or being driven from it by an equal or
nearly equal force; that is to say, it is
enough for such a post to be able to
resist any part of the enemy which might
attempt to carry it without intention of
going farther; for should it be attacked
by a whole column marching upon the
camp, the duty of the piquet is not to
stop the march of that column, but to
give notice of it, and to make its retreat
in good time, without hurry. The officer
commanding will keep his detached posts
on his flanks, halt and fire upon the enemy, whenever the situation of the ground,
                                        or

or the circumstances of the attack, offer him an opportunity, without compromising the safety of his retreat.

Should the piquet be placed in a village where there is a church, it is proper to examine whether the church-yard will not answer for the principal post. That will depend upon the situation of the church, with regard to the roads which lead into the village from the side of the enemy, as well as to the road which leads in a direct line to the camp.

The principal post of the piquet must never be placed so as to leave on its flanks roads by which it can be surprised in the rear, and its retreat cut off: this is the most necessary of all precautions.

Thus, if the church-yard was not situated so as to answer the purpose, the piquet must be placed at the end of the village towards the camp, so as to have before it all the roads that lead to the front and centre of the village. It will then be sufficient to barricade the street where

where it has taken post, either by a parapet of turf, four feet and an half high,
and three or four feet thick at the top,
with the slope necessary to support it ;
or with carts, between the wheels of
which may be passed long pieces of tim-
ber, and such other stuff of the kind
as is to be found in a village. A little
to the rear of that defence, a parapet
of earth and dung, high enough to protect the men against a first fire is to
be quickly raised.

The same must be done at every
entrance of the village from the side of
the enemy, and a small post is to be
placed behind each barricado, propor-
tionable to the breadth of the passage,
reckoning, for a short defence, one man
to every five or six feet.

Care must be taken, besides, to
make for every detached post a foot-path
across the yards and gardens, by which,
in case of attack, each small party may
retire to the principal post, without danger of being cut off by the enemy, should
he

he have penetrated into the village from another quarter.

It is not only necessary to have detached posts at every entrance of the village on the enemy's side, but you must, besides, place double sentries on the outside of the gardens, especially on the salient angles, so as to prevent the enemy from penetrating into the village through the gardens, without being seen in the day-time, and heard or met in the night.

That chain of posts and sentries must be regularly continued through the avenues of the village and the outside of the gardens, in front of the principal post; and on the flanks of the village there ought to be detached posts, corresponding with the other parts of the chain, whether belonging or not to the party which occupies the village.

We must not here confound the measures of defence calculated for a post of resistance, with the simple precautions which belong to a piquet, the

object

object of which is to watch the motions of the enemy, without losing sight of a safe retreat for itself, and the small posts depending upon it.

An officer of light infantry ought to accustom himself to these kinds of combinations, because the details of the chain, with regard to infantry, cannot be regulated by the field-officers of the day, as those of cavalry can; which being employed upon open ground, the general officer who has to form the chain can pass directly from one post to another, placing the vedettes as he rides along. But in an enclosed country, where one cannot pass upon horseback, except by the roads, it would require too much time. In these cases, the field-officers of the day are scarcely able to do more than mark the post of the piquet, and the extent of ground it is to guard; all the rest is left to the care of the officer who commands. If the field-officers of the day were themselves to post all the small detach-

E ments

ments of infantry, it would retard the compleat formation of the chain, and leave the army too long uncovered and exposed to many inconveniences.

The piquet may also be placed in a farm, or detached house. In this case, the commanding officer will examine what is to be done in front of his post.

It would be of advantage to have the view unobstructed for some hundred paces; and to that effect if trees interrupt the sight, he will have them cut down *en abatis*; if hedges, they must be cut down to two or three feet, so as to be able to fire over them.

He will be very attentive not to leave passages open on his flanks, through which his post might be turned and surprized in the rear.

A piquet ought not to shut itself up in a house, yard, or garden, with the intention of defending itself to the last. It may defend its front as long as its flanks are not attacked; but as soon as the enemy offers to surround the post, it must

begin

begin to retire. It is this retreat, which the necessary posts and works are to insure, taking advantage of the ground where it affords passages unseen by the enemy, so as to be able to reach, without being cut off, the road through which the party is to retire upon the camp, or any other post that might have been previously pointed out for that purpose.

The commanding officer will, in the same manner, establish the necessary communications with his posts and sentries, by cutting passages through the hedges within the chain. This precaution will always be found useful, even where there are roads, because the enemy might possibly have made himself master of them, and in that case it is advantageous to have another way.

Those passages and communications must be made and reconnoitred in the day-time, in order to be able to make use of them in the night.

Another road must be prepared for the patroles to go their rounds, from one posts to the other, two or three hundred paces in front of the posts and sentries, through the fields and hedges, on the skirts of woods, or through their interior, as will be found more practicable.

All circumstances and situations are not alike; but some require the greatest precautions and vigilance to avoid a sur- prize. Although a great deal may be said on this subject, it will be found in the practice of it, how difficult it is to say all.

The honours to be paid to general officers or detachments, when they hap- pen to pass near a piquet, will be found in the regulations. To comply with that duty, piquets must keep a sentry on the camp side.

It often happens that the enemy's generals, for the purpose of reconnoitring the ground, approach the chain under the protection of an escort, more or less numerous, and cause the posts to be at- tacked, in order to drive them from the heights, and reconnoitre, themselves, the

the camp and its environs. As soon as the officer commanding the piquet, is informed by his flying sentries that they are approaching, he will instantly send information to the field-officer, under whose orders he is, by one of the orderly light dragoons, who ought always to attend the piquet for that purpose. Should the attack be directed upon the piquet itself, he will do every thing in his power to keep his ground, and prevent the enemy from occupying entirely the height, and accomplishing his end. But if the attack be made upon one of his detached posts, at a certain distance, he will go himself to support it, with a part of his men, and endeavour to dispute the ground inch by inch, in order to gain time, until the assistance, he may reasonably expect, arrive.

If there be any riflemen in the piquet they will be very useful in those circumstances, for the purpose of firing at such persons as might advance singly to reconnoitre.

E 3   Should

Should a trumpeter or drummer, coming from the enemy's camp, present himself, either alone or with an officer, as soon as they are perceived or heard, one of the sentries will go and report the circumstance to the piquet; the remaining sentry will desire them, as soon as they are within hearing, to stop, and turn their faces towards the side they came from. It is to be supposed that the person employed on such a duty, understands the language of the country, or of those he is come to address, either by himself or by an interpreter, that has accompanied him. Till the return of the detached sentry, the remaining one must prevent them from looking towards the camp, and threaten to fire upon them if they do not comply with his directions.

Having received the information, the officer of the piquet will come himself, or send a non-commissioned officer to see what is the matter.

Sometimes it is only a letter or parcel,

cel, which may be delivered without
farther trouble. At other times it will
be an answer, or some other message to
be delivered by an officer. In that case
the officer of the piquet will either go to
his out-posts, and cause a handkerchief
to be tied round the eyes of those who
are entrusted with such letter or message,
or send a non-commissioned officer to
bring them to the piquet with their eyes
covered. He will then report to the
general, and wait for orders to send
those messengers to the camp.

The small detached posts and double sentries will act in the same manner
with deserters presenting themselves upon the chain.

The commander of the small post
will go up to them, and cause them to
deliver their arms before they are allowed to enter the chain. He will, in
the mean-time, send word to the officer
commanding the piquet, informing him
of the circumstance, and of the number
of deserters arrived, that a sufficient
party

party may be sent from the piquet to conduct them thither.

Should a number of deserters present themselves at once, the officer will get his men under arms; and when they arrive at a certain distance from the principal post, he will advance to them and make them halt. If they have not been already disarmed, he will cause them to be so; if they are on horseback, he will make them dismount, and file off, one by one, by the right or left of the post at some distance to the rear, to an appointed place, where they will be kept and guarded by a sufficient number of men.

Those precautions will be more or less extensive, in proportion to the number of deserters who arrive at the same time. Their arms will be deposited with the piquet, and they will be kept in this situation until the arrival of orders from the general, to whom the officer has sent his report, mentioning whatever he may have learned from those deserters;

in

in order that the general may cause
them to be questioned more fully, if
he find any thing interesting in that first
account.

The officer of the piquet will himself carefully examine any person whatever coming from the enemy's side; if tra-
vellers, or people from the country, he
will enquire from whence they come,
where they intend to go, and what kind
of business they have at the camp or in
the neighbourhood.

He will put questions to them upon
what they may know of the enemy and
his situation, what they have seen or met
with in their way, or what they have
heard. From their answers, he will
either stop them or suffer them to proceed,
or send them back, according to the
directions he may have received on the
subject.

Let it be a friend or an enemy's
country, he will take care to behave with
equal kindness to the inhabitants of the
country who bring provisions to camp,
and

and he will not suffer his men to take
any thing from them without payment
or to offer them any insult; and should
he have orders to prevent them from
passing, he will desire them to go back,
in a civil manner, calculated to dispose
them to answer his questions. Important intelligence may sometimes be procured in this manner. However, it may
sometimes be expected that the enemy
will try to introduce spies into the camp,
under colour of bringing provisions; in
such circumstances prudence requires
that men and women of that description
should be closely examined and stopped:
but the officer must not have them searched at his piquet; it will be done at head
quarters, by order of the commander
in chief, if he think it necessary.

Immediately after the first relief,
the officer commanding the piquet will
visit all his posts and sentries, question
them upon their orders, and see whether
the new sentries, know their duty, and
whether

whether the orders have been transmitted to them correctly.

If any of the posts or sentries of the piquet be placed upon a height, from which one can perceive the enemy's camp, and distinguish what is going on in it with a spy glass, the officer of the piquet will often go there, and reconnoitre whether every thing is quiet and in the same situation: whether any troops are entering or going out of it. If he perceive any alteration or movement, it will be his duty to report it immediately to the general.

It is particularly at break of day, before the relief, that the officer will be able to reconnoitre best, whether any change has taken place in the enemy's camp, by comparing its actual situation with that of the preceding evening.

Should it be ascertained in the day-time that the enemy is preparing to move, the officer will immediately report it, and communicate the intelligence to the other piquets on his right and left;
and,

and, holding his piquet in readiness to march, he will continue to watch the motions of the enemy; to follow them as soon as he sees their out-posts falling back to form a rear-guard, as is customary. But in following them he will be very cautious, and not begin an affair without orders or support.

In the mean-time, he will only contrive to reconnoitre which way they direct their march, and continue to report all that he can perceive. To that effect he will content himself with following without hurry, the progressive retreat of the enemy; and causing the country before him to be well searched, for fear of ambuscades, he will successively place himself upon the heights, as they are abandoned by the enemy, and continue to observe him.

In his progress after the enemy, the officer will not fail to inform the general in chief of the different places he may successively occupy, and of whatever intelligence he can procure. It is to be

observed,

observed, that the general must always know where to send his orders to the officer of the piquet, who will employ for that service his orderly hussars or dragoons.

This is, perhaps, the place to remark, that if the country be enclosed and covered the cavalry cannot be employed to advantage in the first line; its assistance will, nevertheless, be of great service on out-posts, either to carry quickly orders or intelligence, or when the ground will now and then permit, to reconnoitre the marches and movements of the enemy more expeditiously than infantry: for that reason, it will always be proper to join a certain proportion of light dragoons to light infantry. In war, no one can hope to act with propriety unless he receive good intelligence.

In the present case, when the enemy is retreating, it may easily be supposed that he leaves concealed behind, in the hedges and covered spots, a certain num-

ber

ber of *chasseurs* or riflemen. Advancing too briskly would only occasion a great loss to little purpose. To dislodge those chasseurs, the best way will be to endeavour to turn them, by penetrating somewhere: afraid of their retreat, they will soon quit the places where they are concealed, and once in march they are no longer dangerous.

If it be our own army which is to raise its camp in the day time, the outposts have no particular movement to make before the time of departure, which the general will always fix himself.

Until that very moment, every thing is to remain in the same situation, and the officer must carefully conceal the order from his men. A little before the time, he will order his riflemen, if he have any, or if he have none, the best marksmen of his detachment, to get ready. Supposing there should be twelve or fourteen, more or less, he will divide them in two parties, and dispose one on the

the right and the other on the left of his ground, and place them at certain distances in the rear of his sentries and small posts, one by one, or two by two, facing such passages as he thinks the enemy will pass through to follow him. Their orders will be to remain concealed till they find an opportunity of firing, and then to fall back upon the party, to load their pieces, and successively take up some other position as before. They must take care to keep a good look out on their right and left, and see which way to take, so as not to be cut off by the enemy.

In case the enemy should not perceive the departure of the army, and remain quiet on his ground, the chasseurs so posted, will then fall back at a slow pace upon the piquet, and occupy of themselves, at certain distances, such positions as will be best calculated for an ambuscade. At the very moment of departure, the officer will get under arms the men who remain in the principal post,

and,

and, in the mean-time, direct a non-commissioned officer to recall all the advanced sentries, and form them into a rear guard, which is to follow them at about two hundred paces distance.

In the mean-time, the small detached posts will be cautioned to fall back upon the principal post; but observing to keep at some distance from it on the flanks, and regulating their march so as not to find themselves cut off from the main body of the detachment by impassable obstacles, such as woods, morasses, rivulets, &c. They are to follow every movement of the main body, that is to say, to march when they march, and halt when they halt. The officer who is supposed to have reconnoitred the ground in the rear, as well as in front, will take care to give them proper directions on that subject.

In case of pursuit, the officer in his retreat will dispose, wherever an opportunity offers, small parties in ambush for the support of the first, in such a manner that

that the enemy may be exposed in his progress to a continual hidden fire, which by forcing him to advance cautiously, will procure to the army the time necessary to pass the defiles, and to make proper dispositions in its retreat.

It is to be remarked, that in all retreats, it is necessary to make a stand at every road, and particularly such as will allow a passage to cavalry and artillery; but it is likewise proper to dispute the possession of the heights, to prevent the enemy from judging from their summits which way the army is marching: and in those circumstances it is easier to dispute the entrance of a practicable road than the possession of a height. To dispute the entrance of a road, and retard the progres of the enemy, it will be sufficient to cut down, so as to fall across the road, the trees which are to be found on the sides of it.

To accomplish this end the officer, having received his orders for the march, will command an intelligent corporal to

take

take with him five or six men; he will form them in the rear of the piquet, and taking care that none but this small party hear him, he will order the corporal to take, in the neighbouring farms and villages, as many men as he can find, with an axe each, and conduct them expeditiously to the road which has been shewn him. The officer will point out to him where he is to begin, it must be at some distance in the rear of the post; he will recommend to the corporal particularly to cut down the trees at the entrance of, and along hollow ways.

From what has been said, it will clearly appear, that an out-post of infantry must be, in proportion to the service, much more numerous than one of cavalry.

If the sentries should, in the daytime, perceive any party of the enemy approaching, one of the two sentries will go to the piquet with that intelligence, as it has been said before. On his receiving that information, the officer will

will get his men under arms, and let them remain covered in their posts till farther orders. He will then take with him a small detachment, and advance to the sentry or small post from which the information came, in order to reconnoitre the enemy's force, whether they are divided in different parties, and advancing directly upon him or upon some other point. Having thus reconnoitred, he will, provided the movement of the enemy should appear to him of any importance, report expeditiously the same, both to the field-officer under whose orders he is, and to the general, in order to be reinforced, or to receive further orders, according to circumstances.

It has been remarked, that whenever there was in front of the chain a river, rivulet, or morass, fordable only at certain passages, the sentinels or vedettes were to be placed opposite to those passages, and were to remain there all night.

In

In such a case, the officer will examine whether circumstances will allow him to make abattis, or raise parapets opposite to those passages, and instead of double sentries, leave there a small post, which thus covered, will be able to make a longer stand, and give time to the officer to come to its support. He will, besides, examine most carefully, whether there is not in front of his post, and in the interval between it and the other posts on his right and left, any other passages, besides those that have been indicated by the people of the country. Should he find any, he will place small posts opposite to them, with the same precautions as the former, that he may not be surprised during the night from ignorance of such passages.

An hour before sun-set, or even sooner, and according to the distance, the officer of the piquet will send the report for the day to the general. He will, for that purpose, employ the non-commissioned officer commanding his
orderly

orderly light dragoons: that non-commissioned officer will bring him back the parole, and the countersign. For the sake of correctness, his report must be made in writing. He will mention in it his own observations, and such information as he may have received from his patroles (if he have been able to send any out), or from the inhabitants of the country.

He must not communicate to any one, either the parole, or any order he may have received. The parole, or counter-sign, is given to sentries at the moment of relief, which takes place every other hour, or every hour, according to the circumstances of the weather, or season of the year. He will send a non-commissioned officer with the parole to each corporal commanding a detached post. If the detached post should be commanded by a serjeant, he will send the corporal to the principal post to get the parole or counter-sign.

With

With regard to the patroles, which one might wish to send out in the day time, in a covered and inclosed country beyond the chain, it supposes, first, that the enemy is at some distance, and that such a patrole is not likely to meet his posts. Should the object be to procure intelligence, and reconnoitre where his nearest posts lay, that service requires separate instruction, which will be found in the section where we treat of recon_noitring.

When the enemy is in sight, in a covered country, it must be expected that the patroles will not be able to advance in the day time, without exposing themselves to the fire of the sentries and chasseurs, disposed in front of the chain. In that case, no information can be procured but from the posts of observation placed upon high ground, from which an extensive prospect may be commanded; or from deserters, spies, or country people.

With

With out-posts, the night is the time of vigilance and precaution; and those vary according to circumstances, which are also of different kinds. Some result from the proximity of the enemy, from his being at a greater or smaller distance, and from his situation, offensive or defensive; others result from the varieties of the ground.

It may happen that posts and sentries, which are sufficient in the day time to discover every thing, and give warning to the principal post, shall not be enough to prevent its being surprized at night, when the enemy can pass between two posts without being seen or heard. It may also happen, that posts of observation, properly placed upon high ground, to discover in the day-time any thing that approaches the chain, would be of no service in the night, from their being at too great a distance from the principal post, and from the road likely to be taken by the enemy, to

surprize

surprize or attack it. This must be ex-. amined.

There is another circumstance which might occasion some alteration in the dispositions for the night : when great darkness prevails, and strong winds, or heavy rains prevent the sentries from hearing the march of troops in their neighbourhood.

The duties of a piquet of infantry at night rest chiefly upon two points :— 1st. To guard the roads by which the enemy might advance upon the camp.*
2d. To guard themselves.

The officer having, early in the course of the day, reconnoitred what alterations it will be necessary to make in the situation of his posts for the night, will be particularly attentive to what regards the defence of the roads within the

---

* In an open country, the roads and fields must be guarded during the night: in an enclosed country it is enough to guard the roads, because a column may undertake to pass through the fields in the day, but not in the night-time

the extent of his post. He will choose
upon those roads the most convenient
spots for making abbatis or cuts; and
without loss of time, he will employ at
that work, as many men of the neighbourhood as can be collected. He will
entrust the care of the work to a serjeant
or corporal, with proper directions how
to have it executed.

The night being come, he will wait
till dark to withdraw such posts as are
placed upon heights, merely for the purpose of observation, and also the sentries
which are to be placed elsewhere. He
will place behind each of the abbatis
and cuts made to defend the roads, nine
or ten men, with orders, in case of attack,
to dispute the road as long as possible,
without compromising their retreat; and
to be sure to know when the enemy endeavours to turn them, they must place
two sentries on the watch, at some distance, one on the right and one on the
left.

When

When they are forced to quit the post, they will retire through the hedges that run along the road, taking care to leave the hedge between them and the enemy, and keep up a brisk firing in their retreat, to show that the enemy is at their heels. However, the officer must direct them to halt, now and then, to see whether they are really pursued and whether it is not a false alarm.

Having thus provided for the defence of the roads leading to the camp, the attention of the officer will turn towards the necessary precautions to guard the piquet against a night surprise.

The new chain, which he will form on the front and the flanks of the principal post, must not be more distant from it than about two hundred paces. This must be particularly regulated by the ground; that is to say, that the distance between the sentries and the piquet must be sufficient to give it time to get under arms before being attacked, without extending

tending the chain more than is necessary for that object. The officer will not neglect, also, to guard his rear; but towards the camp, single sentries will be sufficient.

The sentries must be drawn closer to each other on the circumference of the chain around the principal post, in proportion to the night being more or less dark and stormy. When the night is very dark, and wind and rain might prevent from hearing the march of any thing coming from the enemy, the double sentries will then perform, among themselves, the duty of flying sentries; that is to say, one will remain motionless on the spot marked for their post, while the other will march to the right, till he come near enough to the next post to distinguish the first standing sentry on the right, and then return to his post by the same way.

On the return of the flying sentry, the other, who had remained at the post, will, in his turn, go and reconnoitre the first standing sentry on the left: thus,

suc-

successively and alternately, every sentry will serve as a flying sentry in the interval of two posts. The flying sentry must walk slowly, and stop now and then; and to distinguish better any noise coming from the enemy, he must stoop and approach his ear to the ground as soon as any thing attracts his attention.

By drawing closer at night the chain of sentries around the principal post, and abandoning the heights, which were occupied for the purpose of observing the enemy in the day-time, there will of course be greater or smaller intervals left between two piquets, in which intervals there will be no other standing posts, but such as may have been placed for the defence of the roads, on the right and left of the principal post; and it must be recollected, that the piquet is supposed to have been placed by the field officer who formed the chain, upon the principal road which the enemy would take to march upon the camp on that side of the chain: thus, it might happen, that there

will

will be no other road to be guarded within the post, and that the whole of the duty between two piquets will have to be performed in the night, by lateral patroles. In this case, patroles must succeed to each other continually; and in critical times, the return of one patrole ought not always to be waited for, before another is sent out.

The officer will now have to examine where he will place his own piquet. The piquet is, in general, placed in a thick wood, or a detached farm with yard and garden.

If the piquet be placed in a wood, its skirts must be guarded by sentries, and the officer will fix upon a hollow in the rear of his intended point of defence where a fire may be kept without its being perceived. If there be no such hollow, he will cause one to be made, observing to throw all the earth that is dug out, towards the enemy's side, to form a parapet; and thus, with the help of some boughs and trees which he will

cause

cause to be cut down, he will endeavour to conceal from the enemy the light of his fire in the night. He will take care, that the communication between the fire-place, which must be near his intended point of defence, and the spot where the men fall in, be open, of easy access, and free from bushes.

If the piquet be in a farm or a detached house, the question will be, whether the point of defence shall be chosen outside or inside, including yard and garden.

A house with a yard will but seldom offer a convenient place of defence; because the object of a piquet being chiefly to procure information, the safety and facility of its retreat are more an object of consideration than its means of resistance. But a detachment shut up in a house and yard cannot safely retreat but through the garden; of course the communication, between the house and yard, and the garden, must be very open and easy, and that garden situated pre-

precisely on a certain side of the house, which cannot be always expected. Whenever this is the case, it may be made use of; and as, in general, the point of defence is best placed outside, the officer will choose a spot behind the walls, where he may keep his fire without endangering the building, and without its being seen by the enemy.

He will not suffer his men to sleep in the house. It is in the day that he ought to allow those who are not upon duty, to rest. During the night, the officer will keep one half of those who are not on duty awake round the fire; if he have reasons to be more particularly on his guard, no one must be allowed to sleep.

If, in certain situations, it is not possible to conceal perfectly the fire from the enemy, the officer will order some loose earth or sand, and shovels to be prepared, so as to be able to cover it at the very moment of an alarm.

The officer must visit himself his posts and sentries, at least once during the

the night. Having so heavy a charge upon him, he can expect very little time to rest. In those trying occasions he must recollect his duty to his country, and that an officer, who has once suffered himself to be surprized, can hardly hope to re-establish his personal reputation.

If, during the night, the advanced sentries hear any noise, if it continue, and resemble the noise of troops marching in the neighbourhood, one of them will advance on that side with great caution, and endeavour to find out what it is; and should it prove to be something of this kind, he will return softly and acquaint the piquet.

If they hear, during the night, any thing marching upon them, so near as to be within reach of the voice, one of them will challenge, and if no answer be made, he will fire. If they still continue to approach, the second sentry will repeat the challenge, and fire also; in case of no answer, both will then immediately retire upon the piquet. One-half

of

of the piquet will instantly man the works or defences, and the other half will fall in at the spot appointed for an alarm post.

If the attack begin on the front of the post, such sentries, as are not attacked upon the different points of the circumference of the chain, will retire upon the flanks of the piquet to the places previously pointed out to them by their officer on the right and left: and if they be likewise attacked, they will also retire upon those indicated points on the flanks of the piquet. The officer will always point them out to his men before night, in order that those who are on sentry upon the flanks of the post may know their way and obey the order.

The officer will select two non-commissioned officers, who in case of attack, will go and occupy the places which have been pointed out to them on the flanks of the post. This disposition is of very great consequence; and, during the attack, the officer will turn his atten-

attention particularly to the flanks: the reason is, when the enemy who attacks the post in front, finds it alert, and in a posture of defence, and believes it probably intrenched, he will keep up the fire in front, without advancing; but in the mean-time he will try to pass unperceived by the flanks to take the post in rear, and cut off its retreat. To prevent that manœuvre, detachments must be placed on the flanks, to give notice of it to the officer.

As long as the firing is only in front he may continue his defence; but as soon as he hears the attack begin on his flanks, he will instantly order the men who occupied the defences to leave them and go to the rear, under the protection of those who have remained under arms at the alarm post. Those who have thus quitted the defences will halt, and form at the distance of about a hundred or a hundred and fifty paces, on a spot marked for that purpose in the day-time, and known to all the officers, and non-

com-

commissioned officers under the commander of the piquet. This place must be marked upon the straight line of retreat.

The commanding officer, who till then has answered the fire of the enemy from his post of alarm, will retire in his turn, as soon as he thinks that the post fixed upon in his rear is occupied, in good order, and the arms loaded;* and in this manner his detachment, divided in two parties, will retire alternately upon each other, taking care to choose proper spots for rallying, particularly the entrance of defiles.†

The detachments on the flanks will follow the retreat, observing the progress

of

---

* The whole party must not turn about at once, but leave eight or ten men on the spot as a rear-guard, to follow at about fifty paces distance.

† In order to give time to the detachments on the flanks, in case they should be obliged to retire through the same defile: the defile may also be capable of an obstinate defence, until the support, which might be expected, could arrive.

of the firing, and endeavouring to support themselves upon the same line. The commanding officer will likewise be attentive, not to lose sight of the march of the detachments on his flanks. In proportion to their being more or less pressed by the enemy, he will quicken his retreat, or stand fast, in order to keep as much as possible on the same line with them. He will pursue the same conduct until he receive a reinforcement, or arrive under the protection of the camp. Nothing does more honour to an officer than a retreat made in good order, and adapted to circumstances.

But it will be easily seen, that attention to so many objects cannot be recommended with success, without a perfect knowledge of the ground. The commanding officer, on his way from camp to his post, ought to have followed nearly the same road by which he now retires: he will then have had an opportunity to remark the nature of the country, the situ-

situation of houses and woods, the
cross roads, and the names of the places
where they lead to, and to make, if he can,
a sketch in his pocket-book with a pencil,
or at least to write an itinerary.

The knowledge of the ground being
so essentially necessary to an officer of
light troops, he ought never to pass
through any new part of the theatre of
war, without making such remarks as
circumstances will allow, either in writing, or by taking views of the country.
During the course of a campaign, he will
find, upon many occasions, that his trouble
has not been useless.

If, during the night, any detachment should approach the chain on its
return to camp, the sentries must stop
them at fifty paces from the chain, although they have the countersign, and
one of the sentries will go to the piquet
and report it.

The officer of the piquet will get his
men under arms, and send immediately a
non-commissioned officer and some men,

H   to

to bring the commander of the detachment to the piquet. The detachment is ordered to remain on the outside of the chain till further order.

The commander of the detachment being conducted to the piquet, will be examined carefully by the officer of it, upon all such circumstances as may lead to ascertain who he is, unless he be otherwise personally known to him. This done, the officer of the piquet will give orders to the non-commissioned officer who brought the commander of the detachment to the piquet, to go back to it, which must stand fast until then, on the outside of the chain, and order it to file off, by the flanks of the piquet, towards the camp; the non-commissioned officer will conduct them to the road leading to the camp; the detachment will halt there, and wait the return of its commander, who till that moment will be detained by the officer of the piquet, who will not suffer him to return to his detachment, until the non-commissioned

missioned officer be returned and have made his report.

It may happen, that the returning detachment having been out several days has not got the counter-sign: in that case, the officer of the piquet must examine them still more strictly. However, if no motives of suspicion arise from that examination, he will permit them to proceed to the camp with the same precautions as before.

If any suspicion should result from the examination, the officer of the piquet will cause the detachment to file off, as before-mentioned, to the rear of the piquet. In this situation, he will place some men in front to guard them, retaining their officer with himself. They are not to be allowed in that case to march for the camp, until after break of day.

If a number of deserters present themselves during the night, they are to be stopped at the out-posts, and disarmed, as it has been mentioned before, when speaking of the day-duty. The officer will keep them at some distance in the

rear

rear of the post, under the guard of a few men, and have them brought before him one by one, to interrogate them separately; and if he can collect any interesting intelligence, he will immediately acquaint the general*.

It happens, frequently, that an army decamps silently during the night, without beat of drum or sounding of trumpets.

Whatever be the reason for such a march, it is always concealed from the enemy, and for that purpose the piquets are left at their posts until day break. On such occasions the piquets must be kept very alert, as they may expect to be attacked, should the enemy become

ac-

---

* If any man of the piquet desert during the night, the officer will immediately give another counter-sign to every one of his posts and sentries. As the enemy may make use of this deserter to try to surprize the post, the officer will increase his vigilance, send his lateral patroles more frequently, and keep his men alert. He will acquaint the piquets on his right and left with the desertion, and send them his new counter-sign, in order that the respective patroles and sentries may be on their guard during that night.

acquainted with the departure of the army.

The lateral patroles must be sent out so frequently as to have several in motion at the same time, to prevent spies or deserters from passing through the chain, and giving information to the enemy of what is passing, and also to prevent his patroles from approaching near enough to find it out themselves.

It has been said elsewhere, that during the night, the posts placed upon heights, for the purpose of discovering at a distance, were to be withdrawn at night, but this admits of some exceptions.

If within the extent of the posts of the piquet, there happens to be a height, or other place, from which our camp is to be seen, it has been said, that it must be occupied in the day time. It must likewise remain occupied during the night, and even be then reinforced, if necessary; and the officer will contrive to strengthen it by some sort of work,

H 3   which

which will effectually prevent the enemy
from coming there, in order to observe
the motions of our army.

If there were, besides, in that part
of the chain, a height from which the
enemy's camp could be discovered, it
will be necessary also to keep a post
upon it, during the night. The officer
will recommend to this post to inform
him, without delay, of any alteration
that may be perceived to take place in
the enemy's fires during the night. It
will be proper for him to go and visit
this post himself, from time to time, to
ascertain that nothing remarkable is going on in the camp of the enemy.

When his fires begin to diminish,
and are extinguished sooner than usual,
it is reasonable to suspect that he is making a movement, and when they appear
much greater and stronger than usual, it
may be supposed that he has raised his
camp; as it frequently happens that the
servants, the women, and other followers
of an army, set fire to the straw, and to
the

the sutler's huts, either through neglect
or on purpose, and so it spreads through-
out the camp.

It is the business of those who decamp in the night to take such measures
of police as will prevent it, and to prevent, likewise, the enemy from guessing,
that the camp has been raised, by the
fires going out sooner than usual. It is
customary to leave small detachments of
infantry, to keep them up on the whole
length of the front till about day-break.
As it frequently happens, that what
ought to be done is not done, the
officers commanding piquets must turn
their attention to all those kinds of circumstances and report them.

The two armies are sometimes encamped so near to each other, that if the
leeward piquets pay attention, they may,
in the dead of the night, hear the noise
of troops coming or going, either by the
clashing of arms, the hollow sounds of
voices, or the whips of the artillery
drivers. All unusual noises at night indicate

dicate something. If the noise be heard decreasing, it shews that the enemy is quitting his camp, either the whole army or part of it. If the noise continue and increase, till it cease at once, it may lead to think that fresh troops have entered the camp: if a noise be heard resembling that of pickets driven into the ground, it may be premised that the enemy is working at a battery or intrenchment in the neighbourhood. All those things must be observed during the night by the officers commanding piquets, and each of them must report them to the general as soon as they come to his knowledge.

If an army decamp silently during the night, either to occupy another camp in the rear, or in order to gain an important post, by a march which must be concealed from the enemy, it is customary, as it has been said before to leave the piquets at their posts till daybreak. In this case they must use double

pre-

precautions, and keep their lateral patroles in constant motion.

Among the precautions to be taken for the night, it has been said that the officer ought to reconnoitre in the rear of his piquet, and at some distance from his alarm post, a proper spot to form again, in the event of his being attacked, and obliged to retreat fighting.

In the present case, he will make use of that spot; for if, by break of day, the enemy happen to discover the movement of the army, he will quit his post of alarm and occupy this second position. He will, in the mean-time, send a non-commissioned officer to recall his posts and sentries, who will form the rear-guard with them, and occupy the post of alarm which the officer has just left.

The detached posts will make their retreat upon his flanks, or upon the rear-guard, according to their situation. However this may be, he will take care to cover both his flanks, before he begins

to

to march, by detachments, which must be ordered to observe his march and movements, and preserve, as much as possible, their respective situations and distances.

The officer will not begin to march, until he have received information from the commander of his rear-guard, that all who are under his command have taken their respective posts.

Before recalling his sentries, he will place all the riflemen belonging to his piquet in every favourable situation, and in front of his rear-guard. When those riflemen have fired, they retire upon the rear-guard. Such of the riflemen as have loaded their pieces will take post again in favourable positions, in front of the rear-guard ; that is to say, the rifle-men will serve as a rear-guard to the rear guard.

The detachments on the flanks will be equally covered by riflemen, who will manœuvre in the same manner. For want of riflemen, the officer will pick
out

out the most alert, and the best marksmen of his piquet, to perform that
duty.

Riflemen are principally useful in
retreats; and when, being posted, they
wait for an approaching enemy, if they
be forced to move, and to fire without
being covered, they lose their advantage.

But, with or without riflemen, a
retreat must always be covered by sharp
shooters; and when the dispositions are
well made, and the men practised to this
kind of service, it will be found that
riflemen may be dispensed with.

The officer will fix the moment of
his march, according to the general
movement of the army, and to his being
more or less pressed by the enemy. One
of his principal cares will be to conform,
to what passes on his right and left,
preserving his line with the piquets, near
him, following them with his eyes; attending to the smoke and the report of
their discharges; for it would be very
extra-

extraordinary, that an army, in face of the enemy, could decamp without being pursued more or less.

As he retires, the officer will constantly observe in what manner he is followed. If he can halt a little upon a height, he will observe the force of the enemy, whether it be infantry or cavalry; if this force march upon him, or if another column turn to another side, that he may immediately make his report to the field officer commanding the rear-guard of the army.

If he find on his way a bridge, or other defile, the officer will halt, at the distance of one hundred and fifty or two hundred paces beyond it, forming his men in an arch of a circle, with the concave side towards the defile, so as to concentrate a cross fire upon the passage. Under this protection, his rear-guard will, after passing the defile, form at fifty or sixty paces in his rear, and the sharp shooters will act of themselves according to the nature of the ground.

This

This done, he will leave a small part of the centre of his piquet upon the spot, to follow him at the distance of a hundred paces, and to cover his retreat. He will march with the remainder of his party to form again, at about three hundred paces beyond the spot occupied by his rear-guard; and thus he will continue his march with the same attention and precaution, as well with regard to his own party, as to the general movements of the army.

From what has been said upon the conduct of light infantry in a retreat, it will appear, that to be able to perform with propriety that kind of duty, they must have often practised retreats at exercise. It is not at the moment of actual service that the officer can instruct his men in those numerous details, a great part of which are to be performed individually, and consequently require individual instruction and practice.

The army arrives, sometimes, so late in camp, that night comes on before the

piquets

piquets can be placed. In other circumstances, a piquet receives, during the night, orders to move forward to a certain place or village.

When the place which the officer is ordered to occupy is unknown to him, he will take greater precaution, in proportion to his ignorance of the particulars of the ground. In this perplexity, he will have his first recourse to the map, and in addition to it, he will contrive to procure information from the people of the country.

The first thing will be to procure a guide out of the next houses, and a light, to consult the map. By the answers of the guide to his questions about the nature of the roads and the places they lead to, the officer will be able to judge whether he knows the country, and is able to conduct him; in case of doubt, it is proper to look for one better instructed. Arrived on the spot which he is to occupy, he will again procure a light from some of the country people, and question

tion them as to the villages, roads, woods, and rivulets on the enemy's side; then taking them as guides, make them point out the roads which they have mentioned, in order to place the posts and sentries accordingly.

The officer will mark his alarm post opposite to the principal road he has to defend, conforming to local circumstances.

The place pointed out to him for his piquet may be a single house, the entrance of a village, a wood, a passage, a defile, or a height.

At first he will not have time to think of strengthening his post by art; his first object must be to place his sentries properly, leaving no road unguarded, through which he might be taken unawares. He will occupy the post indicated to him in the manner usual for a night-post; but having lost his connexion with the neighbouring posts, he will take for his flanks and his rear the ame precautions as for his front. He

I 2   will

will give orders to his double sentries to move alternately to each other on their right and left, so that nothing may pass them unperceived. He will keep his whole detachment alert during the remainder of the night. In case of attack, he will conduct himself as has been already mentioned; and to be informed especially of any thing that might come by his rear, he will continually send patroles along the road by which he came, in order to be the better able to direct his retreat, should it become necessary.

At day-break, the officer will reconnoitre the ground which he is to guard in a regular manner, observing what has been said before, with regard to the dispositions calculated for the day-time.

The piquet having kept its ground, and the night having passed quietly, all the dispositions which should be made at its arrival in the day-time, with regard to posts of observation, for procuring intelligence, and other objects detailed before, and which may have been deranged by by the night, must take place at day-break.

A little before day-light, the officer will get his men under arms. This is a necessary precaution, until the ground in front be reconnoitred and searched. Piquets are frequently attacked at that hour. In proceeding to re-establish his day-posts, the officer will observe the following precautions:

He will advance with one-half or two-thirds of his men, according to his numbers and the circumstances of the ground, as far as his night sentries; he will have them relieved, and form them into four, five, or six patroles, of two men each, who will be distributed at the distance of two hundred paces to the front and on the flanks; he will give them orders to search and reconnoitre exactly the bottoms of hollow-ways and ditches, the outsides of hedges, woods, and other cover which might conceal an enemy. He will order two or three small detachments, of five or six men each, to

sup-

support the first, at about two hundred paces in the rear. With the remainder, he will follow slowly at the distance of two or three hundred paces, observing if his small parties search the ground properly.

If the weather be foggy, he will keep the different parts of his disposition less distant from each other. In this case, it is a rule that those who are in front should be able to see each other, the small detachments them, and the main body the detachments.

Having thus satisfied himself that, within his ground, the enemy have not laid any ambush for him during the night, he will replace his sentries and small posts on the spots which they occupied before.

If the fog should continue, and prevent the country from being seen, the sentries will visit each other alternately, as during the night; and if the intervals between the posts were too great, which might probably be the case with day-posts,

posts, he will order continual lateral patroles to be made, to remedy that inconvenience.

It is a general practice to order the
new guards to go at day-break to the support of the old, which they are to relieve
on the chain, because it is, as we have
remarked, the time when piquets are
most commonly attacked, if the enemy
have formed any design against them.

Since it has happened that a piquet
has been carried off violently during the
night, from the neglect of some necessary
precautions, the new guard must approach with care, and halt at about four
hundred paces from the post which they
come to relieve: the small advance guard,
which is to precede them, will move forward and reconnoitre up to the first sentry. This done, the new guard will
advance to the sentry, and remain there
until reconnoitred in its turn by the old
guard, which will get under arms, if not
already so. They will pay to each other
the customary compliments, and the new
guard

guard will take post on the left of the old, fronting the enemy. The officer of the old guard, or piquet, will then communicate all the particulars which concern the post to the relieving officer: in consequence of which, the same number of small posts and sentries will be told off, in the new guard.

The officer and a non-commissioned officer of the new guard, or piquet, conducted by a non-commissioned officer of the old, will go to see the sentries and small posts relieved, and will receive from this non-commisioned officer the information relative to each post.

The old guard will stand fast, until all the posts and sentries be relieved, and the patroles returned; after this, having given all the information in their power, the old guard will march off in good order, for the camp.

If in certain places, the ground permitted the use of a grand guard, or piquet of cavalry, which should occupy in the night a spot in the rear of a piquet of of infantry, as is often the case, it is the duty of the cavalry officer to choose the most convenient place for himself, and take the proper measures for his own security. However, both officers will concert together the duty of their patroles, as well with regard to the time of going out, as to the roads they are to follow, that they may not fall upon each other unawares.

It is still to be observed, that an instruction in writing is little more than an alphabet to serve to form words: it is upon the ground, exposed to the inclemency of the seasons, and in the presence of a crafty and active enemy, that an officer must learn to form his ideas. Nobody can ever undertake to give a compleat and perfect detail of every thing which may happen in war, and to point out all the differences that result from all possible situations.

An officer must not believe, that he has provided against every thing because he has acted in conformity to parts of
this

this instruction. It is in the nature of the ground, in the examination and combination of surrounding circumstances, with the presence and attitude of the enemy, that an officer, by applying to them his own reflections, will discover the manner in which he is to conduct himself.

SEC-

SECTION III.

OF RECONNOITRING.

THERE are different kinds of reconnoitring, according to the object in view.

Before armies are in presence of each other, and when they begin to draw near, they respectively send out reconnoitring parties, which have orders to push forward, until they discover the out-posts of the enemy; and, if possible, the body of his army. This kind of reconnoitring may occasion numerous affairs and skirmishes.

When armies are in presence of each other, reconnoitring is made to discover how they are encamped, what extent of ground they occupy, and as much as possible, the strong and weak part of their position. In this second kind of reconnoitring, every sort of engagement must

be

be avoided, as much as possible. However, there are cases where the purpose cannot be accomplished without attacking and driving in some of the enemy's outposts.

Reconnoitring takes place in the day and in the night, upon the front and flanks of the enemy, whenever it is the object to discover his movements, and to know when he marches in part or in totality, to what side, and with what number and kind of troops.

We reconnoitre round the flanks of the enemy, towards his rear, in order to become acquainted with the ground by which we would march to occupy a position to turn him, and thereby oblige him to change his camp, or make a disadvantageous movement.

That kind of reconnoitring, which concerns the figure and nature of the ground, the facility or difficulty of marches, the merits of a position, &c. &c. are, in general, made by an officer of the staff, with an escort proportionate to the

prox-

proximity of the enemy, and the extent of the commission he is to execute. In this case, the commander of the escort is under the orders of the staff-officer, who points out to him, from place to place, the road he intends to take; but the dispositions and precautions, for the safety of the escort, belong, nevertheless, to its commander.

No reconnoitring of any of those kinds is ever undertaken, without an express order from the general, accompanied with instructions which fix positively the nature and object of it.

What constitutes particularly those different kinds of reconnoitring, is the difference of precaution in the march, in proportion to the object. To describe those precautions, it becomes necessary to apply them to the objects. It is to be recollected, however, that we do not treat of reconnoitring in general, but of that kind only, which takes place in countries where cavalry alone cannot be employed without inconvenience.

K

§ 1.

§ 1.

If it be supposed, that the enemy might be met at the distance of eight or ten miles, more or less, by following a certain road; and if the object be to ascertain how near he has pushed his advance-guard, a detachment of a hundred infantry and twenty dragoons*, in proportion to the disposable force, and other circumstances, will be ordered for that purpose.

Seven or eight light dragoons, with a corporal, will form the van, at about seven or eight hundred paces from the detachment. At that distance from the infantry, the corporal will detach before him two dragoons, who will march by files, sixty or eighty paces asunder, taking care not to lose sight of each other; approaching for that purpose when the road is winding, and keeping

at

---

* This detachment ought to have bread, at least, for two days, and the horses a sufficient quantity of oats for the same period.

at a greater distance when it is straight; but always so, that the first dragoon may be seen by the second, and the second by the corporal.

The dragoons at the head of the infantry must march at a regular pace, at the rate of seventy paces a minute at most; but if they stop frequently, they must make it up again.

At every cross road, the corporal of the advance-guard will detach a dragoon, who will advance five or six hundred paces at a fast trot, more or less, until he can have an extensive prospect, and examine whether there be any troops in march. This dragoon will question such of the people of the country as he meets with, to know whether they have seen an enemy in the neighbourhood.

The first and second dragoons, who form the van of the advance-guard, must frequently look behind, to see that they are followed. If the first see that he is not followed by the second, he must halt; and if the second do not see the corporal following him, he must also halt. The second dragoon will march again, the moment he perceives that the corporal is advancing, and upon the movement of the second, the first will also recommence his march.

As often as the corporal detaches a dragoon to his flank, he will halt until he return, and make his report. Should he have seen or heard any thing, the corporal will acquaint the commander of the detachment. The corporal will stop, and send to the commander of the detachment every person whom he meets, either going towards the enemy, or coming from that quarter.

The commander of the detachment having, as has been said, suffered the cavalry to take the start of him by seven or eight hundred paces, will form an advance guard of twelve or fifteen light infantry, at about two hundred paces distance from him. He will also detach, on each of his flanks, seven or eight men and a corporal, to march parallel to the road,

road, at about three hundred paces from it.

Whenever there is a wood or farm on the sides of the road, at the distance of about three or four hundred paces at most (woods at a greater distance are not to be searched), the commander of the detachment will halt, until his flankers have searched the wood, to a certain depth*, and likewise ascertained that no enemy be concealed in the farm.

The advance-guard of cavalry can only reconnoitre upon the road: they cannot search woods and houses without dismounting.

If the road go through a wood, the corporal who leads the van will halt a little from it, till the infantry flankers be arrived. They will search the wood to the distance of three hundred paces from the road on each side, and then the two dragoons and the corporal will move on,

---

* Two or three hundred infantry can be easily concealed, in the barns and stabling of a large farm, when the country people favour them.

fol-

followed by the detachment at two hundred paces distance. At the end of the wood, the corporal will trot gently on until he regain his former distance in front. The remainder of the dragoons will form the rear-guard of the detachment, at three hundred paces from it.

Persons stopped by the dragoons of the advance-guard, going towards the enemy, may be allowed to pursue their way, by marching between the rear of the infantry, and the dragoons of the rear-guard.

If among those coming from the enemy's side, there should be any who have given interesting intelligence, if the commander think it worth while, he may either send this person with a dragoon to head quarters, in order to be more particularly interrogated, or he may dispatch a dragoon with a written report of it. This, and several other things, may be thought too trifling to be mentioned, but this book is not written for those who have already seen actual

tual service, but for those who require to be taught every thing.

In proportion as you approach the place where you expect to find the enemy, the precautions must increase; and the woods on the flanks must be searched with more care, and to a greater distance.

It may happen, that the enemy, informed of the march of the detachment by the people of the country, his spies, or his out-posts, will conceal some infantry in a wood, at five or six hundred paces from the side of the road, with orders to remain concealed until the detachment be passed by seven or eight hundred paces, and then to take possession of the road and neighbouring hedges, at the back of the detachment, which would be thus attacked in front, and soon after in the rear. A detachment must be particularly careful not to fall into such a snare; and, for that reason,

son, the rear-guard of dragoons must remain at a greater distance from the detachment, in proportion to its getting nearer the enemy, and likewise make a short halt, fronting the woods on the flanks, in order to see that no concealed party be rushing out of it to seize upon the road.

Should it be the case, the dragoons of the rear-guard will instantly acquaint the infantry, and take post so as to preserve the power of retreating, without going too far from the detachment. However, the dragoons are by no means to draw near the infantry, but the infantry is to draw near them.

On these occasions, the dragoons are to make use of their carabines to fire at the enemy, over and through the hedges; not for the effect of their fire, but as a warning to the infantry; and an inducement to the enemy to stop and return it. Should he do so, it will give

to

to the infantry time to return, and make,
on the opposite side of the road, dispositions both for defence and retreat*.

At

---

* If at that distance the detachment have to cross
a river, or considerable stream, which cannot be passed
except by a bridge, upon the road of the detachment,
the commander will leave there some foot soldiers, with
orders to intrench themselves, and two dragoons, one of
whom will take post on the other side of the bridge,
and the other on this side of it, in the rear of the infantry. This small detachment must be the stronger,
if there be no other bridge in the neighbourhood.
This precaution must be taken, in case that the enemy
having got wind of the march of the detachment,
have come by some cross road to seize upon the
bridge; and should this be actually the case, the
dragoon in front of the bridge will gallop by the same
road which the detachment followed, and acquaint
the commander: the dragoon in the rear will ride back
to camp, and report the circumstances. If at this moment the commander of the detachment be near the
enemy, he will do every thing in his power to avoid
an engagement; and should he have learned from his
guides, that there is another bridge on the same river,
more distant from the enemy, he will direct his retreat
upon it; but if there be none within a reasonable
distance, he must expeditiously return, and drive back
the enemy, whatever may be his force, before he has
made himself master of the passage.

At the first shot or notice, the dragoons of the advance-guard are to fall back upon the detachment, and proceed straight to the rear. By proceeding in that manner, sooner or later, one is sure to discover the enemy.

At the first sight or notice of him, the advance-guard of cavalry will fall back upon the rear-guard of the same arm. Their duty will, from that period, be only to observe and report whatever might come by the rear on the flanks of the infantry, taking care not to allow their retreat to be cut off, without keeping, however, at a greater distance from the detachment than is necessary for this purpose, and preserving carefully the power of communicating with the camp.

The commander of the detachment will be provided with the best map of the country that he can procure, and a good spy-glass. He will change his guides at every village; but if among the first there should be one possessed of a more general

general knowledge of the country, he will do well to keep him.

Every time that the detachment arrives upon a height or commanding ground of any kind, he will get from his guides the names of all the objects in sight, and write them in his pocket-book: he will also examine, whether any of those objects are marked in his map, to know better where he is, and what road he must take.

He will question his guides with regard to the roads, rivers, and streams in the neighbourhood; and particularly of such rivers and streams as his detachment has to cross; whether they are passed upon bridges, and where these are; or if they are fordable all the way, or only at certain places, and where these places are. He must be acquainted with those particulars, in case he should, by some accident, be obliged to retreat by another road.

Detachments of this sort must always be provided with two guides; one

to

to be on horseback, if possible, with the corporal who leads the van, and the other will either ride or walk by the side of the commander, at the head of the infantry.

As soon as the enemy is discovered, or has been announced, the commander will order his detachment to halt. He will examine his situation and the circumstances of the ground, and expeditiously make his dispositions accordingly.

He will find himself, either upon low ground, commanded within musquet shot; or upon a commanding height; or in a plain.

If he find himself on ground commanded within musquet shot, he must hasten to quit it, either by advancing, or retiring to the nearest height in his rear, or to the sides of the road in a narrow valley, or only to one side, if the valley be wide, to gain the nearest high ground.

Here the variety of circumstances may give rise to a number of dispositions, too long to be detailed; but although different,

different, according to the ground, they nevertheless refer to the same object.

The first thing for the officer is to put himself in a situation to judge of what is advancing, whether it be the advance-guard of an army, or only a detachment. It is for this first reason, that he ought to gain the nearest height, whether in front, rear, or flank. Arrived there, he will immediately divide and place his men, so as to cover and protect his flanks, and not allow himself to be turned, while some of the enemy's sharp shooters attempt to divert his attention to the front. His dispositions must be calculated to advance; or, according to circumstances, to retire, by simply facing about, as soon as he has obtained the information wanted, or that it appears there is danger of being turned by a superior force. He will take care to have always one division ready to support the part which falls back, and however hotly he may be pursued, he must perform his retreat in this successive manner.

L   **The**

The first plate gives an idea of a disposition in a plain, or other ground, not commanded. The main body of the detachment is divided in two nearly equal parts, A and B; the remainder is seen man by man, and the cavalry closes the march at about three hundred paces.

If there be on the right or left, a height, valley, or wood, under cover of which the enemy could turn the detachments, the officer will send there a corporal and four men detached from A or B, to reconnoitre as far as a thousand paces; their firing will be a signal that the enemy is marching to that quarter. As to them, whether the detachment retire or advance, they will follow the progress of its march by the same flank, and fire, as a signal, whenever they perceive the enemy.

The proper manœuvres for those separated parties, to line hedges, and support those who successively fall back in a retreat, ought to be practised as an exercise for the instruction of light infantry;

for,

for, without it, it would be very difficult for the commander of a detachment, at the moment of acting, to teach his men how to occupy the ground, and preserve their distance before the enemy.

These little manœuvres require the more practice, because there is nothing fixed in them. If there be a rivulet in front or in flank, a hollow way, &c. &c. changes must be made in consequence, which must be altered again, when these no longer exist. For this reason, light infantry ought to be often practised to sham retreats on different kinds of ground, in order that every one, officer and soldier, may know how to act in every variety of ground, and in whatever manner he may be attacked and pursued.

The commander of the detachment having fixed upon the spot upon which he is to wait for the enemy, the skirmishers, C, D, E, F, G, will conceal themselves, as much as possible, under cover of the neighbouring hedges, preserving nearly their general and respective distances.

The

The small detachments, H, Y, K, will take post, so as to be able to support their skirmishers in front; that is to say, will be prepared to take up a new position behind the hedges, as soon as the enemy has dislodged them. The dislodged skirmishers will fall back upon the detachment which is ready to support. The serjeant or corporal commanding it, will point out to them other hedges, under the cover of which they will make another stand; and so on.

If the ground be open, the officer will easily see what he is to do; but if the prospect be circumscribed on every side, his principal attention will be turned to his flanks. He will, as before-mentioned, send on each side, beyond the line of his flankers, a corporal and four men, taken from the body of the detachment, in order to reconnoitre the ground on the right and left, at about seven or eight hundred paces. They will keep at distance, and give notice of the approach of an enemy by firing, taking care,

care, at the same time, to follow every movement of the detachment.

As long as the firing is respectively confined to the front, there will be no necessity for taking new measures; the skirmishers will continue to fall back slowly, and not before they are forced to do so. But if the enemy be numerous, things will not remain long in this situation. The small parties on the flanks will soon be heard to fire. The first shots must only attract the attention of the officer; but as soon as the firing appears to increase on that quarter, no time is to be lost. He will cause the retreat to be sounded. We use the expression, *sound the retreat*, because we think the bugle preferable to the drum, for light troops.

The retreat having sounded, the light infantry will understand, that they are not to wait until they be forced to fall back, but are to retire upon a serjeant or corporal in rear of their second line, as soon as they see it has taken post.

L 3. This

This non-commissioned officer will post them again, and so on successively.

The first part, A, of the main body, will support the skirmishers; and should the enemy press too hard upon either flank, this division, A, will march to its support, the whole or a part, according to the violence of the attack; and the officer will be principally attentive to prevent the enemy from outflanking him. To prevent this, he will cause the retreat in quick time to be sounded, in order that the skirmishers on the front and flanks may fall back briskly, without endeavouring to stop the enemy, who might seek to amuse them in order to turn them.

The small parties of observation detached beyond the flanks will, by degrees, draw nearer the detachment, as soon as they hear the retreat sounded, and quicken their march when they hear the second retreat.

The commander of the detachment having ordered the retreat to sound, must recall

recall to his mind the circumstances of the ground by which he is to retire.

With the second division, B, of the main body, and the cavalry of his advance-guard, he must march, as expeditiously as possible, to some favorable spot on his intended way; such as the summit of a height, not commanded, or an enclosure of trees; the hedges or walls of a garden, the windows and doors of a house; taking care, if he occupy a house ever so well situated, to open a retreat by the back part, and not suffer himself to be shut up in it.

He will take advantage of thickets and hedges, from which he can scour the road, and establish cross fires upon it, so as to stop the enemy, and gain time to put his detachment in order again, should the pursuit have created a little confusion. If he think himself in danger he will dispatch a dragoon to camp for succour.

Having rallied near him those who have retired from the fire, he will form them

them again, according to the ground, but so as to preserve always a body of reserve : he will then order the bugle to sound the retreat, and in the mean time he will, with the reserve, march briskly to the rear, and occupy the nearest favorable position, to stop the enemy again, and repeat the same manœuvre as long as he is pursued, or till he receive reinforcement, or arrive under the protection of the out-posts of the army.

If the detachment, in its retreat, have to cross a river or rivulet, unpassable except upon bridges, we have already pointed out what would be the first precaution to be taken, in order to secure a free passage: we shall now examine, whether the ground admits of defending the bridge, until the arrival of succour.

If there be no height apposite, commanding the passage within musquet shot, the officer may undertake to defend the bridge, under the cover of the cuts or traverses which the detachment, left for that purpose, have already made.

Should

Should the bridge be commanded, so as not to allow of a good defence, the officer will leave only the detachment which had the guard of it before, to dispute the passage as long as practicable, and he will resume his manœuvres of retreat, as soon as the superior fire of the enemy obliges the small detachment to abandon the defence of the bridge. If in its retreat the detachment have to pass through open ground, such as a common, the commanding officer will order his cavalry to go first, and take post, as indicated in the second figure.

The infantry having passed, and the sharp shooters remaining posted, as in C, D, *fig.* 2, the cavalry will begin its retreat by the wing which, from the circumstances of the ground, is most exposed to the enemy's fire. If one be not more so than the other, it is immaterial which retires first: in the second figure it ought to be the right wing. From this moment, the retreat will continue to the camp in the order described.

We

We have said before, that as soon as the officer has ascertained the presence of the enemy, his first care should be to form his detachment, and place it so as to observe what is coming upon him, without compromising himself; and so as to be able to march forward, or retire, according to circumstances, without changing his order of battle.

If the enemy, informed of the presence of the detachment, march briskly to attack it, it is a certain sign that you have fallen in with the head of his advance-gurad, or some strong detachment, conscious of a considerable corps being at hand to support it : in this case, there is no choice, and we have already pointed out how to perform the retreat.

If the enemy do not make an immediate attack, but, on the contrary, be observed not to advance, but to take on his side the same precautions, it indicates a common detachment, without even support at hand.

If while you cause his front and
flanks

flanks to be reconnoitred carefully and cautiously, for fear of an ambuscade, he contents himself with keeping up a more or less brisk fire, from his skirmishers; from this conduct the officer will conclude, that to act with propriety he must try to make some prisoners, from whom positive information may be obtained. For that purpose, the officer will cause the bugle to sound the march, in order to engage the enemy's skirmishers, and drive them from their post upon their main body.

But before marching to attack, the officer will examine, attentively, the nature of the enemy's post, and its environs. He will also question again his guides, upon whatever may serve to regulate his conduct. He must observe to question his guides before the firing begins, at least before they can fear its effects; otherwise they are, in general, frightened, and incapable of speaking to the purpose.

So situated, and having procured the infor-

information he wanted, the officer may dismiss his guides; but if he think he may still want them, he will have them guarded by the cavalry in the rear, and even tied, if they should appear inclined to run away.

Having taken the resolution to attack, in order to procure a few prisoners, the officer will examine by what part of the ground he can best approach the enemy's flank, concealing his strength. He will then cause the charge to be sounded; that is to say, a sound of the bugle horn, previously agreed upon, as a signal for the skirmishers to engage and advance upon the enemy; observing, however, that they are not to come absolutely to close quarters, but to approach as near as they can, by firing.

The officer having fixed the manner of his attack, will take with him the whole, or part of his reserve A, B, *fig.* 1, according as he thinks necessary. He will be preceded by some skirmishers, who are to search carefully the hedges, thickets,

thickets, and hollow ways, where the enemy might lay in ambuscade.

If the skirmishers discover an ambuscade, the officer will march straight upon it, in good order, without firing or stopping an instant*. By an abrupt charge he will put them to flight, and if on this occasion he can take a few prisoners, his end will be accomplished, and he will, in consequence, cause the *partial* retreat to be sounded; that is to say, the retreat for that part of the detachment under his immediate command †.

If the enemy retire so precipitately that no prisoners can be made from those who formed the ambuscade, the officer will

---

* If the skirmishers discover an ambuscade they are to retire immediately upon the detachment, without firing, in order to acquaint, verbally, the officer of the circumstance.

† The bugle-horn must have different sounds for retreating; one intended only for the skirmishers, another for the divisions which are actually manœuvring, and a third one for the general retreat. Those different signals must be made known to the troops at exercise.

M

will detach twelve or fifteen men to attack in flank the skirmishers of the enemy, who may be still engaged in front with those of the detachment.

It is evident, that in order to take prisoners, you must rush on, without firing at those you wish to take. Such prisoners as you may happen to take in this manner, must be brought instantly before the commanding officer, who will then order the partial retreat to be sounded, for that part of the detachment which is actually manœuvring with himself. And having, as expeditiously as possible, collected his party, he will form in the same order as before the attack.

The prisoners having been disarmed, will be sent with a small escort to the rear, where the cavalry, in proportion to the number of prisoners, will join a few dragoons to the infantry, and proceed, without delay, with them to the camp. This done, the officer will order

order the general retreat to sound, and fall back directly upon the army.

If, in the progress of his attack, the officer does not meet with any ambuscade, he will continue to move on, so as to out-flank the enemy, and thus cut off the greater part of his skirmishers. If he fall upon any of the enemy's small covering parties, he must strictly forbid his skirmishers from returning their fire: they must march forward silently, till he command them to halt. All those orders may be communicated with the bugle-horn.

Having penetrated as far as he intends, the officer will fall briskly upon the flank of the enemy with one-half or two-thirds of his men, observing always to keep a little reserve with himself, ready to parry accidents, and to give support where it may be wanted. Having made a few prisoners, he will retire upon the army, as mentioned before.

§ 2.

When armies are to be reconnoitred, in order to ascertain how they are encamped, and what are the advantages or defects of their positions, it is done by the general in chief, or by some principal officers of the staff, under the protection of an escort, the strength of which is regulated according to that of the enemy's out-posts, and in consequence of the nature of the ground they occupy.

The troops of the escort have nothing to do in conducting the business; they receive their orders, in that respect, from the person who is to reconnoitre. He will indicate the road they are to follow, and point out the enemy's guards, which are to be driven back, in order to be able to ascertain the strength and extent of the encampment, which is to be reconnoitred.

Sometimes a reconnoitring of this kind is made on the same day, and

at

at the same time, upon the whole front
of the enemy's line; sometimes upon a
part, and successively upon another, as
it is the general's intention to hasten, or
to defer his attack.

It is no part of the duty of light
infantry to make dispositions for those
kinds of reconnoitring, it concerns the
officers of the staff; for the principal object is to reconnoitre the ground, and the
manner in which it is occupied by the
enemy. In this case, the officers of
light infantry have only to cause the
ground to be well searched in front of
the march, and on the flank nearest the
enemy, taking care that nothing can
come, by surprise, upon the person who
reconnoitres; and for that purpose the
commander of the escort, with a part of
his men, will place himself between the
general or officer who reconnoitres, and
the rangers and flankers, so as to follow
and cover their march, keeping himself
at a proper distance to receive orders,

and

and communicate them to the detachments by means of the bugle-horn.

The out-posts of the enemy being
generally placed on the heights which
cover the avenues to his camp, it happens very often, that to reconnoitre his
position, it becomes necessary to drive
back the chain of out-posts upon the
camp: in that case, the general (as has
been expressed before) will determine
the place, the time, and disposition of the
attack.

However, it will, perhaps, be not
useless to observe, that such kinds of
attacks must be brisk, and supported
with great spirit by a number of troops,
superior to those you wish to dislodge.
The sharp shooters at the head, must be
numerous, and closely supported by the
small detachments, upon which they
are to fall back, if forced so to do;
and that first line is to be supported by
as many troops in reserve as are judged
necessary. The enemy is to be approached as near as possible without
firing;

firing; the whole party is then to advance at the quickest pace, and from all quarters. It will continue briskly to push back the enemy, until the situation and extent of his camp, can be distinctly seen : then the commander of the escort may take the general's orders to push on, or halt and rally, which will give the general time to make his observations.

Having finished the reconnoitring, it is customary for the general and his suite to retire first and separately, with a small escort of cavalry. As soon as he has got to a sufficient distance, the commander of the light infantry will order the retreat to be sounded, and in every thing else he will act in conformity to what has been said concerning the dispositions for retreats.

§ 3.

When two armies, of nearly equal strength, are in presence of each other, and are advantageously encamped, without either of them having strong reasons

to

to come to a decisive action, it happens
that the one which has the greater quantity of forage, either green or dry,
obliges necessarily the other to decamp.
The party which is obliged to move first
is always exposed to some disadvantage;
and it is to be informed of the moment
and direction of its march, that small
parties are sent to watch its movements,
and report immediately every occur-
rence.

It sometimes happens, that an inferior army remains encamped in a chosen
and intrenched position, in order to force
the enemy to make lateral movements, or
to attack, under great disadvantage, the
position in which it has intrenched itself.
In such cases, it is to be expected that
the enemy, profiting by his superiority,
will either turn the intrenched position,
so as totally to prevent, or throw obstacles in the way of foraging parties and
convoys; or he will intrench his own
camp, which will enable him to spare
large detachments to act separately, and
take

take advantage of his superiority in some way or other.

The choice of a position to wait for the enemy, without inconveniency, and with an inferior force, depends upon a variety of circumstances and considerations, which form an essential part of the science of generals, which we do not treat of here. It will, then, be sufficient to remark in this place, that the position not being in a defile, whose flanks are inaccessible to a great distance; in all other situations there are movements by which the enemy will oblige him, who is on the defensive, to quit his position, either because it it has been turned, or that the enemy by having divided his army, has an opportunity to form enterprises, which it becomes absolutely necessary for him to oppose.

To judge what is to be done in a situation which requires the greatest talents and most scientific manœuvres, it is evident, that a general ought to be thoroughly acquainted with what the enemy

enemy intends to do, at the very moment he puts his army in motion, in order to regulate his own measures with security.

To bring the general accurate intelligence of this nature, it is necessary that the officer of light troops, employed for that purpose, should possess some knowledge and talents. It is not that they are required to reconnoitre and guess at the designs of the enemy; they are only to report what actually takes place; whether the enemy remain in his camp, or raise it; in what number he marches, and in what direction. The general will judge what may be the object of the movement of his adversary.

Therefore, without interfering with the science of generals, it is, however, necessary that an officer should have a just idea of the consequence of the reports which he is charged to make in these delicate occasions, in order to be positive only, with regard to what he has seen and verified himself: mentioning,

ing, also, what he may have heard, but
without conjectures or reflections of his
own.

Detachments of this kind consist
seldom of more than twenty or thirty of
the best chasseurs, and five or six dragoons. It would be difficult to find every
where proper places to conceal a greater
number, nor would it be easy to provide
for their subsistence. The detachment
must carry meat and forage for three
days, and the meat must be cooked, as
no fire is to be made use of on such an
occasion, either in the day or night. It
is usual to relieve detachments every
three days, on account of the fatigue
they are exposed to, and that the little
rest which they can take, is always interrupted.

It is different with the officer, who has
been chosen for this duty. Whatever personal fatigue he may have undergone, he
ought to remain with the new detachment;
for the more he has seen of the ground,
the more he will be able to perform the
duty

duty required; which a new come-officer could not do with so much facility and advantage.

The part of the ground upon which the detachment is to try to place itself will be on the flanks of the enemy, by turning his out-posts, and approaching his rear, which usually is less guarded; arriving thus on the back of the wing, which is to be observed, and drawing as near as circumstances will allow.

Although it be always of service to question the inhabitants of the country, upon the nature of the ground upon which it is intended to act, in the present case it is not to be done. The officer must regulate his conduct by his own *coup d'œil*; and, in a march of this kind, avoid, as much as possible, to be met by any of them.

However, if, previous to his approaching the flank of the enemy, the officer meet with a village, out of the reach of the enemy's patroles, it will be proper for him to speak with some intelligent people

in

in it, who may be acquainted with that part where he intends to go.

If the officer see upon his map any river or rivulet in his way, he must inquire whether they be fordable or not, the nature of their banks, whether flat or perpendicular, and deep; whether there be morasses in the neighbourhood of them; whether the country be intersected by vallies, woods, heights, &c. &c.

These questions must not be made directly concerning the place in view; but the officer will inform himself, at the same time, about different other places, in order to puzzle those whom he is questioning, and to prevent them from guessing at his motives, or at the direction of his march.

If he cannot, by those means, procure satisfactory intelligence, his best way will be to conduct himself from place to place, by his *coup d'œil*, taking care not to quit one, until he have caused another to be reconnoitred. That is to say, if it be proposed to cross a valley,

the

the officer will detach one or two men,
who profiting by every advantage which
the ground offers to conceal their march,
will examine whether any unpassable
morass or stream pass through the bottom of it. Arrived there, if the officer
intend to gain a neighbouring height, he
will likewise detach one man, or two,
to examine the summit and reverse, and
report what they have seen.

During this time, the detachment
must remain concealed behind a hedge or
thicket; observing to place a few men in
front, to warn them of whatever may approach. By moving on, in this cautious
manner, from place to place, the officer
will, at last, succeed in occupying a point
upon the enemy's flank, at a distance
from his out-posts, and out of the track
of his patroles.

Having reached this spot in the day-
time, the officer will look round for some
height, covered with trees and hedges,
and distant from any habitation; or what
would still be preferable, for a wood,

so

so situated, as to command a view of whatever goes out of the enemy's camp by that flank. He will make such re-marks, as will enable him to find his way there, during the night; for he must take care to avoid houses and roads: and if he suppose that it will be too difficult for his dragoons to accompany him in this march, he will point out to them a corner of a wood, or other place in the neighbourhood, where they are to take post with some light infantry, to assist them in guarding themselves during the night. He will give them orders to hold themselves in readiness to receive and carry full speed any intelligence to the general.

The officer will take care that his men be provided with a few whistles, to call and give answer. By means of these whistles, which can be heard at a considerable distance, especially in the night-time; the light infantry man, sent to carry the dispatches to the dragoons, will whistle, in case he should not find

them

them at the place pointed out to him, as it may happen that, for greater safety, they have removed to another: but, in this case, they will leave one of the light infantry concealed in a hedge, or upon a tree, to answer the whistle, and carry the report to the new place of concealment of the dragoons. In this case, the dragoons serve only to forward reports with more expedition; consequently, if the country be such as not to allow of this service being performed by cavalry, the officer will leave small posts of three light infantry each, at the distance of about three miles asunder, upon the whole line of communication, and thus the reports, conveyed from post to post, at full speed, will arrive as expeditiously as if they had been carried by cavalry.

Every one of the small posts must be provided with a whistle, that they may readily find each other by day and night, and the detachment must be stronger, in proportion to the number of small posts neces-

necessary to forward the ordinary, as well as extraordinary, reports: for ordinary reports must be made every evening, whether any thing have happened or not, and extraordinary are sent to communicate any new occurrences which may be seen or learnt.

There must be different manners of whistling, distinguished by numbers. The number indicating the whistle for each day will be given to the light infantry with the parole.

Although whistling may be of great service to light infantry on many occasions, it is, however, subject to inconveniencies in others. For this reason, the officer will allow or forbid the use of it, according as, from the situation of his posts, it may be heard or not by the enemy; and, whenever whistling is forbidden, clapping the hands, or some other signal must be substituted.

If the light infantry man who carries the report do not find the first post where

it

it ought to have been, he will go on to the second, to the third, and if he can find none, to head quarters.

The eye of the officer having fixed upon the post of observation, as it has been already mentioned, he will proceed to it at night, with great silence, observing to avoid villages and houses. Arrived there, the detachment must not be allowed to make a fire, nor to converse, nor to make any kind of noise. Those only who are not on sentry, and who can neither be seen nor heard in striking fire, can be allowed to smoke.

At day-break, the officer will place sentinels behind trees and hedges, in the front and round his post, so as to see clearly whatever passes in the neighbourhood, or approaches him; and if the country on the enemy's flank cannot be well seen from the hedges, the officer will cause a few of the most active men to climb the trees. They will inform him of every thing they see, communicating

cating it by men who are to remain at the foot of the trees; they must observe to speak low.

When the importance of the report renders it necessary, the officer will go himself to the place from which the movements of the enemy can be seen, and endeavour, with his spy-glass, to re-connoitre the circumstances. He will mark, in his pocket-book, the result of his observations, and the hour he made them; and if a march, or other important movement of the enemy, be likely to happen, he will immediately dispatch an active man, with a written report of the circumstances, who will proceed by the rear of the post, taking care not to be seen or met by any one, marching with the greatest precaution, hiding himself in corn fields, or hedges, when there is danger of being seen, and remaining till dark, if necessary; and if he see it impossible for him to escape being taken, he must tear the report to pieces, and scatter them, and give himself up as a

deserter,

deserter, afraid of being pursued. Should the report be of very great consequence, it will be incumbent upon the officer to dispatch another, half an hour or an hour after the first, for fear it should have been taken; he will, in this second report, mention what has happened, or what he has seen since the departure of the first; and if he have to announce the march of the whole army of the enemy he will continue to dispatch a messenger every hour, more or less, with the report of every thing which occurs. If nothing essential happen during the day, the officer will not send his ordinary report till evening.

In the choice of the post, the officer must take care not to advance too far towards the rear of the enemy, when it may be supposed that he is more likely to march by his flank than to the rear. Those circumstances must be mentioned in the instructions given to the officer.

When two armies, of nearly equal force, occupy positions, and forage becomes

comes scarce with one of them, it may be easily foreseen that this army will soon retire. In this case, the officer of the detachment must endeavour to penetrate farther to the rear, the better to ascertain the movement of the baggage and heavy artillery, which, in a march to the rear, will take the lead of the column.

If, on the contrary, the enemy be on the offensive, it will be sufficient to take post on the line of prolongation of one of his wings, in order to discover when he begins to move by the flank. In this case, his advance-guard will lead, followed by the columns without baggage.

If, in a movement of this kind, the officer have penetrated too far to the rear of the enemy, it is evident, that he will be under the necessity of quitting the post instantly; for if he remain concealed in it, he will not be able to forward intelligence, as the communication with his camp is cut off by this movement of the enemy. For these reasons, the choice of such a post is a difficult matter.

matter. It is also requisite to take post, so as to be able to reconnoitre the movements, which frequently take place during the night.

If the object of the detachment be to observe a retrograde movement of the enemy, the officer will probably see, towards evening, some of the baggage assemble in the rear of their camp, in order to march off, followed shortly after by the heavy artillery. Upon these signs, and as the darkness increases, he will detach a few men, who passing along from hedge to hedge, will ascertain whether it be really the whole of the baggage of the army which is in march. This they will be able to do from the incessant noise of the whips of the artillery and waggon drivers; and the march of columns is known by the buzzing noise of the infantry, and the clashing of arms. They will easily distinguish the march of cavalry by approaching the ear to the ground.

But when a detachment is sent to observe

observe the motions of an army acting
upon the offensive, it may happen, that
the officer will discover no appearances
of a march during the day, and will hear
the retreat beat as usual, without any in-
dication of a movement about to take
place ; excepting, perhaps, a certain
number of detachments sent out to fur-
nish flankers for the march, and also to
prevent spies from passing and giving intelligence of it : this must be attended to.
It may then happen, that about two
hours after retreat, the enemy's army will
get under arms silently, and their advance-guard move on, while the tents
are struck, and the baggage takes a different road, covered by the march of the
army.

To be able to judge, before hand,
of those kinds of movements, and prepare
the means of discovering them, the officer
of a detachment must have acquired so
accurate a knowledge of the country, as
to be able to place his light infantry in
the vicinity of the roads which the enemy
might

might take during the night; especially as a night-march, in a covered country, cannot be performed across the fields without previous preparations, and that the columns of the enemy must unavoidably follow the roads which are in the direction of his intended movements.

If the officer remark that the enemy send pioneers to construct bridges, repair roads, or open passages across the fields, he may conclude that he intends to march from the right or left of his camp, and he will not fail to send in an exact account of such facts, as soon as they come to his knowledge.

By observing great precaution not to be discovered, either by the enemy or by the country people, when they are unfavourable, the officer may perhaps be able to continue day and night in the same post for some time; however, prudence requires that he should fix his eye upon another post, and have it reconnoitred by one of the most intelligent of his light infantry, in order to have it in his

his power to retire there, as soon as he may think himself discovered, or in danger of being so. Having thus two posts, he can occupy that which is more favourably situated to discover the enemy's camp in the day, and retire to the other at night.

If the officer judge, by the direction of his patroles, that the enemy, informed of his presence, is searching for him, prudence requires he should cause another post to be reconnoitred, to which he can retire, in case of necessity, and have it in his power, at the same time, to change his post at night, taking care to plant his sentries, so as to prevent a surprize.

In such circumstances, sentries must not challenge, nor ask for the parole, nor fire if no answer be given ; but by whistling, or clapping hands, acquaint the post that somebody is coming that way, and retire in the mean-time, in order to make their escape with the others, by that side where no whistling has been heard, and which, of course, is supposed clear. At

all

all events, the detachment must not fire : its safety does not depend, in this situation, upon the use of its arms, but in its vigilance and legs. It is easy to conceive that it would be impossible to continue in the neighbourhood after firing, and giving the alarm.

The officer will always quit the night-post a little before day-break, so as not to be seen on his way to the post of observation. If on the preceding day he has remarked any thing which leads him to conclude that he is suspected or discovered, he will, in this case, send only two or three of his men to the post of observation. They will conceal themselves in trees and hedges, and continue to observe, while the officer will retire with the remaining part of his detachment to a covered spot in the neighbourhood, where he is not likely to attract the notice of the enemy. If the detached men report any thing which renders it necessary, he will go to them himself, and ascertain the truth,

It

It may happen, as has been remarked, that the officer may be obliged to change his post several times during the night, from information, or other circumstances which might render that precaution necessary, to avoid falling into the hands of the enemy. By changing frequently, he will puzzle those who are in search of him, and thus be able to make his escape before he can be surrounded.

If the detachment be in want of any thing, the officer will take care that nothing shall be asked or taken in the houses and villages near him, but he will send a man or two to procure it at a greater distance.

As the principal object of a commission of this kind is by means of the knowledge of the country, to observe silently the movements of the enemy, the officer, or his men, are on no account ever to think of taking prisoners; and, above all, the officer must exert himself to prevent his detached men from

taking

taking any thing from houses or gardens, for by giving cause of complaint to the inhabitants, the detachment would soon be pursued and found out.

If, in spite of every precaution taken to avoid being seen by the inhabitants in an enemy's country, any of them should directly pass through the day-post of the detachment, the advanced sentries concealed in the hedges will suffer him to go on, until they can seize him from behind, and prevent him from escaping the same way he came. One of the men will then conduct him, in a civil manner, to a place at some distance from the detachment, where he must be detained till night. The men must not abuse him in any way; but tell him that they are forced to use these precautions, because they are deserters, pursued by their officers, and waiting for the night to continue their journey.

When dark, the detachment will proceed to its night-post; and when the two soldiers, who have the care of this

man,

man, think that it is a sufficient distance, they will conduct their prisoner to a road in a different direction, ordering him to pursue his journey. The two soldiers will wait till he has got to a certain distance, and proceed to the place appointed to join their detachment. At their approach, they will clap their hands as a signal, which will be answered in the same manner; for it might be dangerous to use the whistle when near the enemy, as observed before.

A commission of this kind is attended with so many different dangers, that it is necessary for the officer to be able to depend entirely upon the fidelity of his men; and he will take care, in assembling the detachment, to advise those among them who have money, to leave it with the pay-master or commander of the regiment: because it frequently happens, that men, otherwise faithful and brave, will not behave as they ought, when exposed to lose their money.

The last precaution to be taken is,

to

to fix upon, and point out to all the men a rallying place of rendezvous, in case the detachment, being surrounded, should be under the necessity of dispersing and saving themselves individually. This, however, is not to be done without necessity, and the express command of the officer, given by a signal of the whistle.

When the detachment is to be relieved, the officer will send to camp one of the men best acquainted with the country, to conduct the new detachment to the night-post which has been fixed upon for the purpose. The light infantry-man, who conducts the new detachment, will endeavor to arrive in the vicinity of the post about dusk, and will remain concealed, until it be sufficiently dark to prevent the enemy from perceiving the march of the detachment. When he approaches near the rendezvous, he will give the signal agreed upon by clapping the hands, which will be answered if the old detachment be there; but this cannot be be certain, since from hour to hour circumstances may occur, to oblige the officer to change his post: in this case, the relieving detachment must be conducted to the second night post agreed upon.

The old detachment will take the best road to camp, and the officer will, if his health permit, continue on this fatiguing service, until recalled by the general, or by the favourable opportunity of an important report.

§ 4.

There is still another kind of reconnoitring, with regard to the movements which may be undertaken by the enemy: it is when armies are encamped so near to each other, that the one might attack the other by surprize, during the night.

In such circumstances, the patroles of both armies meet frequently in the night, fire at each other, and alarm both camps: but both armies soon get

used

used to this kind of noise, and pay little attention to it.

If one of the two generals have formed a plan of a night attack, he will order his patroles to alarm the enemy's piquets during several nights, on the whole extent of its front, and to retire always at day-break; expecting, by these means, to decoy the enemy into a false security; and such alarms being now little attended to, to approach with his columns, so as to fall briskly, by day-light, upon the principal batteries, and to force the camp, before all the enemy's troops have time to get under arms, and occupy the field of battle in good order. The probability of success in attacks of this kind, depends chiefly upon the ground. Sometimes they are rendered practicable by a wood of easy access, or in which several roads, leading to the enemy's camp, are in the hands of him who attacks, or by a number of small successive valleys, which shorten

shorten the prospect and favour a surprize.

To prevent this, it is customary to
send out, at dusk, a few detachments of a
certain force, which take post a little
beyond the piquets, and opposite to
such passages as the enemy is most likely
to come through. Those detachments
must remain concealed and silent until
attacked, and will then march forward
in close order, and make a brisk charge
upon whatever may be in their way,
until they arrive at the place to which it
may be supposed that the enemy's column
have advanced in the dark.

If the officer judge, by the noise
which denotes the march of troops, that
he has got near a column, he will halt
his detachment, and order a heavy dis-
charge in the direction of the noise.

However contrary the orders may
be, it is difficult to prevent men from
firing during the night, when they think
they are attacked, and the fire of the
detachment will soon be returned in a
manner

manner which will betray the enemy's column.

When the officer has satisfactorily reconnoitred the march of the enemy, he will retire by the same way he came, ordering his men to keep well together, and march in close order, until the detachment reach the piquets, or the camp, according to circumstances.

In his retreat, the officer will use the precaution to halt now and then, to put his men in order, and also to reconnoitre and listen around him, directing his march accordingly. The best way to fall in with the enemy's detachments is always to march in compact order, and charge briskly whatever may be in the way.

If the officer, having penetrated to a certain distance meet with no large body of troops, nor discover the march of a column, he will, for awhile, keep concealed and silent on the same spot, and return to camp a little before daybreak.

Having,

Having, by his march to the front, ascertained the real situation of things, the officer will, in case he should have met with or reconnoitred an enemy's column, carry immediately that information to the general; but if he have only met with an ordinary patrole, which his superior force enabled him easily to put to flight, he will not make his report before he has seen his detachment back into camp.

§ 5,

Among the different kinds of reconnoitring, by day or night, it frequently happens that a detachment is sent out, to ascertain whether the enemy does or does not occupy such or such a place, and in the first case, what are his force and disposition.

The commander of the detachment will begin, as mentioned before, by fixing the number of his advance-guard and flankers, according to the force of his detachment. This applies to day,

as

as well as night-marches. If the reconnoitring be to take place at night, the
advance-guard and flankers must keep
near enough to the detachment to prevent any danger of a separation and of
losing their way. The greatest silence
must be expressly commanded, and no
one be allowed to smoke, strike fire, or
take dogs along with him, as is often the
case among soldiers.

At the meeting of several cross-
roads, the serjeant or corporal commanding the advance-guard, must be
sure to leave a man, to shew the detachment which to take.

If great darkness prevail, it will be
necessary to place some men between
the advance-guard and the detachment,
forming a sort of chain, by which means
the march of the advance-guard can be followed exactly by the detachment, which
will march and halt in the same time
and in the same manner as its advance-
guard.

The two men who are to march in
front

front of the advance-guard will halt
frequently and listen; and, if they
hear any noise, stoop and place their
ears to the ground, in order to ascertain
better of what kind it is, from what number of men it appears to proceed, and at
what distance.

If the dogs should bark near the
road, it may be occasioned by the approach of the detachment; but it may
also be in consequence of something
coming from another quarter. In this
uncertainty, the detachment will halt,
and the officer will send two or three of
the most intelligent men, who will advance cautiously towards the place where
the barking is heard.

If it be a village or house, which
they have approached without being discovered, they will examine whether there
is any light, make towards the side where
it shows, and endeavour to arrive at it
by creeping through the hedges, yards,
and gardens, avoiding, as much as possible, the paths by which they would be

likely

likely to be met. Arrived at the house, they will endeavour to find out whether there are any soldiers in it ; and by approaching gently to the window where the light appears, endeavour to see what is passing in the room.

If they see in it none but people of the house, one of the men will rap gently at the window, and request to speak with the master of it, or any other person in the room. He will beg of him to speak low, and not to be afraid of him, as he only wishes to be informed whether any troops be in the village or in the neighbourhood, of what kind, and in what number, as far as comes within his knowledge. Having received the information wanted, they will retire silently, and report to the officer, who will act accordingly : retire, if he find that his object is accomplished ; or proceed farther, if he wish to learn more.

By acting in this manner, it is not unlikely but the officer may find an opportunity of surprizing the enemy, in

some

some farm or village, where he is badly guarded. It is the officer's business, to consider his means of success, the safety of his retreat, and above all, what are his orders and instructions.

When a fire is seen at a distance, the detachment will halt, and send a party to reconnoitre, for whose use it is kept, whether for an enemy, country carters, or shepherds. If it be an enemy, the officer leaving his detachment on the spot where it has halted, will go himself, and observe, at a proper distance, those who are near the fire. From their number, situation, and manner of guarding themselves, he will judge what is proper to do.

If he think it right to attack them, he will cause a proportion of his men to gain, silently, their rear, with orders to advance and attack on their side, as soon as they hear the first shots ; and when the officer thinks that they have arrived on their ground, he will fall abruptly upon the enemy in front, who, thus surprized, will

im-

immediately give way. The fugitives will be met by that part of the detachment which has turned to their rear, and prisoners can easily be taken. Having secured a few, the officer will cause the recall to be whistled, and comply with his instructions.

If it be ascertained that the people by the fire are country people without arms, one of the men sent for intelligence will approach them, and ask them such questions as are necessary. The officer, having been informed of the answers, will, if he think it of service, go himself to them for particulars. If the commander of the detachment be not sufficiently acquainted with the country to dispense with a guide, he must not fail to take one with him.

When a detachment is sent from camp, the staff-officers, who are to march it off, and give it the necessary instructions, will also provide it with a guide. But it may happen, that a detachment, not being sent off from camp, the officer must

must provide a guide himself. With regard to the conduct to be observed with them, the officer must consider whether he is or not in an enemy's country. When there are reasons for not trusting them, precautions must be taken; in this case, it would be proper to chuse for guide a housekeeper or his son, and threaten to burn his house as a punishment, if he betray the detachment or lead it astray.

If one were obliged, for want of a housekeeper or proprietor, to employ a servant or day-labourer, who could run away without having any bad consequences to fear, it will be necessary to have him tied by the middle of the body, and placed under the care of a soldier, who will pass the other end of the cord round his own waist. He must also be informed, that the soldier is ordered to kill him, if he attempt to run away or mislead the detachment; but that he will be well used and rewarded, if he lead them right, and give every information in his power to the officer, concerning

the

the roads, and places where the enemy might be met with or heard of.

The men employed for night detachments ought to be such of the light infantry as speak best the language of the country. This will favor the deception, and enable them, during the night, to pass for friends with the natives, and thus procure much useful information.

If the detachment be to pass a bridge, upon a stream that is not any where fordable, or a dyke, across ponds, or morasses, or any other kind of defile, the detachment will halt, until it be ascertained that no enemy is in the neighbourhood on either side.

To guard the passage, a few men will be left, who will fire, to alarm the detachment, if the enemy should appear, and offer to make himself master of the passage.

If those men can procure a little straw and dry wood, it will be placed a little behind the bridge, so as to be easily lighted

lighted and seen at a great distance, to serve as a signal, that the enemy is in possession of the bridge. But as it is not always practicable to find the necessary materials, and a proper situation for this signal to be seen at a distance from the road which the detachment is following, it will be easiest to give information of the approach of the enemy by a few shots; and should the wind blow in a contrary direction, so as to make it probable that the report could not be heard, it will be proper to leave some men at certain distances to repeat the signal. It is to be observed, that they are to fire in a perpendicular direction, in order that the flash may be seen during the night, in case the report should not be heard. The intermediate posts will look constantly towards the side of the signal. These precautions will be necessary only if the detachment intend to return by the same road: in the contrary case they will not.

If the road which the detachment is

to

to follow happen to draw near the front and out-posts of the enemy, the officer will take the precaution to keep on the flank which is next the enemy, at about four or five hundred paces distance from him, two or three small parties of five or six men each, in order that in case any corps of the enemy should be in movement in the neighbourhood, these small parties may stop them, until the detachment have gained time to get out of their way, and continue its march. As for the small parties themselves, they must, in every case, whether they are attacked or not, return to the camp, without attempting to join the detachment, which would oblige it to wait, and occasion loss of time.

When a commission of this kind is to be executed in the day-time, the officer will, as usual, detach a few men to his front and flanks; these will search the woods and the bottom of vallies, and ascend cautiously the summits of the heights on the road which the detachment

is

is to follow, to discover if an enemy appear in the neighbourhood. While this reconnoitring is going on, the detachment will halt in some covered spot, and not proceed before the flankers have made their report. If any patroles or detachments of the enemy be perceived, the officer will judge by the course they hold, whether he can remain where he is, or must retire elsewhere.

The principal object for him is to avoid being seen or met, before he arrive at the place mentioned in his instruction. Once there, he will look out for some coppice, or other cover, sufficient to conceal him, and plant sentries around it to prevent surprise. After this, he will send to different quarters the men who speak most fluently the language of the country. These will conceal themselves near the roads leading to the place in question, and avoiding carefully to be seen by such people as may be going to it, will approach with civility those who are coming from it, and ask them such questions

as

as were previously ordered by the officer. In order to discover the truth, they may give themselves out for deserters, and say that, being afraid to be retaken, they wish to know whether there are any troops or not in the place those people come from, and whether the road is clear.

If, by chance, the person they speak to offered to go back, they will oppose it; and the surest way will be, in this case, to seize him, threatening to kill him if he offer to cry out, or resist in the least degree. In this manner they will conduct him, a prisoner, to the detachment, to be kept in custody there, until the moment of departure. Even the country-people in the fields may be questioned with certain precautions.

To ascertain the truth, it will be necessary to speak to several persons coming from that place, and from the conformity of their report, the officer will be enabled to judge of the degree of credit they may deserve.

In some cases it will be sufficient to send

send a spy to the place, who will introduce himself under pretence of selling or buying provisions: but in other cases, it will be safer and more expeditious to send a detachment, which will act as has been mentioned.

At other times, the object is to ascertain promptly, whether the enemy occupy or not this or that village, market-town, wood, or height; this is called direct reconnoitring, and the dispositions for the march, attack, and retreat, laid down in the first paragraph, will answer equally well here.

It may also happen, that the detachment sent for intelligence, is ordered to try to make a few prisoners, in order to ascertain how things are: but the manner of taking prisoners by surprise, belongs to the section upon *Ambuscades*, and therefore shall not be mentioned here.

With regard to that kind of reconnoitring which concerns the direction of a march, and the nature of camps and positions,

positions, it is the business, as has been said before, of the staff-officers. Light infantry have nothing to do with it, but as an escort, and their duty, as such, has been already explained.

SECTION IV.

Conduct of an Officer ordered to make Prisoners.

At certain moments, in certain circumstances,* when from want of spies information cannot be procured of what is passing in the enemy's camp, it is customary to send at the same time, and to different quarters, small detachments, for the purpose of making prisoners.

These prisoners, coming from different places, and belonging to different corps, will report what is going on, where they were individually; and by comparing their different accounts, more or less light will be thrown upon the situation and projects of the enemy.

Q   The

---

* Sometimes the chain of the enemy is so well placed and guarded, that the spies who may be in his camp cannot get out of it, but by going a round-about way, and, of course, bring none but stale news.

The manner of executing commissions of this nature depends chiefly upon the particular ingenuity and cunning of the person entrusted with them, and also upon the varieties of the ground upon which he is to try to surprise the enemy, either in the day or in the night.

It is to be remarked, that those commissions are performed with more facility by cavalry, where the ground allows them to act; and that they are proportionably more arduous for infantry, when the ground prevents the assistance of cavalry.

However, the first care of the officer entrusted with such a commission, before he form his plan, must be to acquire some knowledge of the roads which the enemy's patroles are used to follow, and how his chain is situated with regard to him.

To obtain this, and approach the circumference occupied by the enemy's piquets and grand guards, he will act as reported in the section *Of Reconnoitring*. To prevent his being seen or discovered, he he will avoid high roads, and even common roads and villages, creep from wood to wood, and from height to height, under the cover of hedges, and never quit a place of concealment, before he have examined and caused to be reconnoitred where and how he can find another.

In order to facilitate his march, and the reconnoitring necessary to its success, the officer will previously procure, in some village of the neighbourhood, about half a dozen pieces of coarse cloth, like sacks, or old blankets, wide enough to cover a man from the head to the waist. He may get these articles from the municipal officers of the place.

It is to be remarked, that whatever be the colour of the uniform, the light infantry will always be liable to be discriminated at a great distance by those among the enemy who have a good sight. The head dress and cloathing of a soldier are too easily distinguished from the dress of a peasant; but, with the assistance of the

above-

above-mentioned sacks, which will cover their head and shoulders, they will be able to advance without being known, taking care to leave their firelocks with the detachment, or to carry them so as not to be seen.

The detachment having thus reached the spot where it is to halt, the officer will cast his eyes around, to see if he cannot find a height, from which he may observe, himself, the situation of the enemy's posts, the march of his detachments and patroles. He will then conceal his detachment in the most covered place, planting a sufficient number of sentries for its security. This done, he will cover himself with one of the sacks, and advance as near the enemy's out-posts as practicable, without creating suspicion.

Taking post in a place whence he can see at a distance, without being himself discovered, he will wait with patience, until he see some detachment or patrole of the enemy in motion; and he

will,

will, by the means of his spy-glass, endeavour to find out the disposition of the
enemy's posts and sentries.

As it is the usual practice for piquets
to relieve their sentries every second hour,
he will pay attention to the march of the
relief, which will give him an idea of the
disposition of this part of their chain ;
and if he discover a post which is not
relieved from the piquet, he will conclude, that it is occupied only by a few
men and a corporal, as a post of observation. From those different circumstances he will judge what the ground
and the force of his detachment permit
him to undertake.

Infantry against infantry, and, in a
covered country, the way to take prisoners will always be to attack in the rear
those you wish to seize upon : and, for
that purpose, the enemy's chain must be
penetrated somewhere, before the post
meant to be attacked can be turned.

A double attack will, in general, be

found

found advantageous, because it engages and divides the attention of the enemy. The false attack will be made in front, corresponding exactly with the real one, which is to penetrate rapidly through the chain, turn, and fall upon the rear.

As the plan comprehends two attacks, so there must be two commanders to conduct them separately; and it must be considered whether it be to take place in the day-time, or whether it would be better to wait for the night. The officer having determined the time and manner of his attack, will call the non-commissioned officer to whom he intends to entrust the command of part of it. (He might have previously taken this non-commissioned officer along with him, under the disguise spoken of;) he will point out to him his way from field to field, from hedge to hedge, from tree to tree, and from height to height, giving him instructions for his conduct by day or night. To be well seconded, the

officer

officer must explain himself clearly, and agree perfectly with him upon what is to be done.

The part of the detachment which is to act in front having approached the enemy's post, as near as can be done, by creeping under the cover of hedges, will halt, where it is deemed expedient, this party being intended only to divert, in one way or other, the attention of the enemy's sentries, so as to prevent them from reconnoitring the march of the other part of the detachment: for this end, various means and stratagems may be employed, either by day or night.

According to circumstances, two men may be appointed to present themselves before the enemy's post, and make signals that they have something to say. If the sentry fire at them, they will make no answer, but repeat the same signals, and approach within hearing; they will halt at that distance, saying that they are deserters; they will wait apparently for the corporal, or other commander of the post,

to

to come forward, to reconnoitre and disarm them, as is customary. But before he approach too near, they will begin some discourse, drawing back by degrees, and saying, that they hope they will not be ill-used, nor forced to enlist, and other things of the same kind, if they can speak the language of the enemy. In this manner they will endeavour to decoy the corporal from his post, and detain him by their discourse, so as to give time to those who are advancing by the flanks to cut him off from his retreat, and seize him, and any others who may have accompanied him.

The officer may disguise entirely as a peasant one of his men, who can speak the language of the country, giving him a basket full of provisions, or of fowls tied by the legs, which he will take care to make cry. Every thing being prepared, the man with the basket will advance towards the enemy's sentry, and beg leave to pass with the provisions to camp, offering to make him a present of part,

part, if he will allow him to proceed with the rest.

The soldiers of the post will undoubtedly approach and examine the bearer of the basket, who, seeming to be frightened, will draw back instead of advancing. Having got to a little distance he will lay down his basket, preparing to run away. If the men of the post pursue him, or stop to take his basket, they will either fall into the ambuscade, or be caught by those who have marched to turn them.

There are many little stratagems of this kind, the success of which depends on the person who employs them; and it must be remarked, that without some sort of stratagem, it is not easy to approach sufficiently near to the front of a post of infantry to make prisoners: because, if it be a strong post, the attack will be repulsed, and if it be weak, it will make off too suddenly to admit of making prisoners.

A com-

A commission of this sort is much more easily executed by cavalry, where it can act. They have only to conceal themselves, and fall rapidly upon an enemy of the same arm. If they put them to flight, they are sure of making prisoners; because the good horses of the party, which attacks and pursues the other, must overtake the bad ones of the vanquished and flying party.

It may, however, happen, that while the officer is making his observations, he may perceive a detachment of the enemy or some of his patroles, proceeding along the chain. If the party approach towards him, it will be his business to examine whether the ground presents a proper situation for an ambuscade, and in what manner the party conducts its march. If it march with precaution, it is probable that it would discover the ambuscade, and it might be dangerous to be engaged in that situation with a stronger party; but if he

remark

remark that their rangers content themselves with searching the ground, to a small distance from the road, he may plan his attack accordingly, and remain concealed until the detachment have passed. He will immediately send seven or eight men to attack the head of it by a road which he will point out, and, as soon as he hears them actually engaged, he will march on briskly, and fall upon their rear-guard. This he will be able to perform while the detachment is engaged in the pursuit of those who have made the attack in front, and are ordered to direct their retreat straight upon the army.

When you see a patrole going out or returning, you may guess that, in a few hours after, another will pass the same way. In this case, the officer will also have to fix upon, and contrive the means of arriving at a place where he may lay concealed, and cut off the patrole from its out-posts. To do it with more certainty, he may divide his detachment

tachment into three or four small parties, which all rising up by signal, at the same instant, will prevent the escape of the patrole, whichever way it may turn.

If none of those contrivances appear practicable in the day-time, the officer will wait till night, and approach that piquet of the enemy, which appears to him to be the easiest to be turned. He will remark, previously, the ground, so as to be able to direct his way in the dark ; and it will be his business, according to the nature of the ground, to fix upon his real and false attacks, and to employ some stratagem, which will enable him to approach and deceive the enemy's sentries, and even surprize the post if possible.

It happens, sometimes, that the enemy will have day-posts, which are withdrawn at night. This kind of information may be got from the people of the country, if they are upon your side; or it may be guessed at, when one of the enemy's posts is seen placed upon an

insu-

insulated height, separated from their chain by a rivulet or valley.

If the enemy occupy such a post day and night, an attempt may be made to carry it by turning it during the night; and if the post be occupied only during the day, the officer will take proper measures to go there at night, and post himself, so as to be able to surround those who come as usual to occupy it at break of day.

When an officer receives an order to make prisoners, in order to obtain some information, which cannot be procured from common spies, it implies that the information wanted refers to a particular object. For instance, to know whether a reinforcement is arrived in the enemy's camp, or whether troops have left it for some particular destination; whether heavy baggage and artillery have been sent to the rear, &c. &c.

The officer must be previously acquainted with the questions he is to put to his prisoners, and will interrogate

R   them

them accordingly. He may promise to set them at liberty if they speak truth, and threaten to shoot them if they try to deceive him. But it very often happens that they know nothing, or at least nothing to be depended upon. It is only by comparing reports from different quarters, that an inference can be drawn : for this reason, the officer must content himself to report simply the intelligence which he has collected, by interrogating his prisoners singly, and not add any thing positively, unless he have an opportunity to ascertain himself the truth of the prisoner's reports.

If, during an expedition of this kind, the detachment should want provisions, the officer will use the precautions mentioned in the Section of *Reconnoitring*, as well in procuring them, as in every thing which has not been repeated here.

THE DUTY OF LIGHT INFANTRY IN BATTLES.

Of the Duty of Light Infantry, during the Dispositions which precede Battles; in Battle; and after the Battle.

THE duty of light infantry in battle, cannot form the subject of a particular chapter, unless the theatre of war be supposed to be in a country where the nature of the ground prevents cavalry and infantry from marching and acting in line. In such a country, light infantry is employed to advantage, in covering the front of attacks, in connecting them together by occupying intervals, and in protecting their flanks; by gaining the hedges, woods, and uneven ground, of which the enemy might avail himself.

R 2

What

What we shall say, therefore, upon this subject, will be confined to the advantage which may be derived from light infantry in battles, where the ground requires the assistance of such troops.

Battles may be considered under two different heads, attack and defence: and although two armies may march reciprocally to attack each other, yet it will generally happen that the nature of the ground will determine one of them to halt, and to receive the attack, in order to secure some local advantage ; or one of them having completed its dispositions sooner than the other, will be able to march first to the attack. Therefore, every thing which concerns battles, is always to be referred to the dispositions for attack and defence.

SECTION I.

Of the Duty of Light Infantry in the Dispositions for Attack, previous to a Battle.

THE duty which light infantry has perform, when the army marches against the enemy, is to cover the march, by the dispositions and precautions already related.

When the heads of the columns halt, either to deploy, in conformity to the dispositions of the attack, or to form, till farther orders, into close columns of battalions, it is the business of the light infantry to cover the deployment of the columns, by occupying the houses, hedges, hollow ways, and thickets, in front, to the distance of about six or seven hundred paces, according to the proximity of the first batteries of the enemy, and the situation of his out-posts. In

the.

the execution of this duty it sometimes becomes necessary to dislodge the enemy from places where his out-posts have intrenched themselves, so as to be capable of some resistance. .

Such posts, however, are not to be attacked without orders from the general. Not that the attack can be dispensed with, but that measures for supporting it may be necessary, without which the troops might be exposed to an unequal conflict. The light infantry will, therefore, continue to skirmish before such intrenched points, until orders for advancing be received.

In the dispositions for attack which concern light infantry, and in fact at all times, we have already observed, that the main-body, in the rear of the skirmishers, was to be divided into two parties, at two hundred paces from each other; it is with the advanced party that the enemy's right or left is to be turned.

It is the business of the officer to reconnoitre the ground, and fix upon the

the flank of the enemy, which he ought to turn, in order to fall upon his rear.

Those who intrench themselves will, in general, endeavour to occupy the highest ground, either to prevent being commanded, or in order to see better what advances against them; but wherever there are heights there are also vallies, and it seldom happens that the fire from a height can equally protect all the inequalities of the ground.

The *coup d'œil* of the officer who leads a detachment to the attack, ought to be well practised. He ought to be able, at once, to fix upon the road, most likely to conduct to success, with least bloodshed.

The sides of heights and vallies, are either concave or convex : the convex swell out towards the middle, which forms a double declivity. Concave sides of moderate elevations offer no shelter from fire, and must be passed at such a distance, as will prevent it from taking

its

its full effect; but, when very high, there is generally some part about the middle, more or less extensive, where the fire from the summit cannot reach. It is only necessary to arrive there, and continuing to move along on a line, which cannot be seen from the intrenchments on the top, thus endeavour to turn them.

If this line, however, be flanked by a work of the enemy, it is requisite to begin by carrying it. If the enemy have intrenched this part, its low and grazing fire will greatly augment the danger of the attack.

A thick hedge in low ground, a hollow way, a ravine, a defile, which the enemy has neglected to occupy, will often be sufficient to enable light infantry to turn a post, which could not have been attacked in front, without occasioning a severe loss.

The variety of those local circumstances must be carefully examined. The ma-

manœuvre peculiar to light infantry is, in all attacks, to penetrate by turning the post.

Light infantry, acting in the midst of hedges, woods, and upon rough ground, can never attempt, in close order, a regular attack to overturn whatever presents itself : these are actions reserved for whole battalions of grenadiers, whose bravery is not always successful, though supported by a powerful artillery.

It is particularly by contriving to penetrate, by stealth, through neglected parts, (and in a covered country it is almost impossible there should be none), that light infantry can attempt to attack fortified posts, and thus carry them from behind : and it is in the same manner, that it can also contribute, materially, to the success of regular attacks, made in front by the troops of the line.

When the columns have finished their deployment, and the dispositions of the general are complete, the light infantry

will

will retire at a given signal,* either upon the flank of the attacks, or into the intervals of the line. These places must be particularly pointed out to them by the staff-officers, whose duty it is to communicate to the troops the orders of the generals.

---

* This signal may be given by the bugles of the light infantry, who may be distributed from distance to distance, along the front of the line; the call being sounded at the center, the order will be communicated in an instant to the whole.

SEC-

SECTION II.

Of the Duty of Light Infantry in the Dispositions for Defence, previous to a Buttle.

WE have said, that when points capable of resisting occur in front of the enemy's field of battle, it is the business of light infantry to take every advantage of the ground, and of the negligence of the enemy, to endeavour to turn and to attack them in rear.

The officer entrusted with the defence of such a post, in front of the field of battle, or of a certain extent of ground on the flanks of a post, must carefully examine every covered spot, in front and on the sides of it, within musquet shot; must reconnoitre every passage by which the enemy could turn his post, and soon oblige him to cease from defending it in front.

At

At certain distances, along his whole line of defence, he ought to conceal two, three, or four sharp-shooters, according to the importance of the point, with orders to watch over that part of it. With the disposable part of his remaining force he will form two reserves, one for the defence of the right flank, and the other for that of the left.

Each reserve ought to be placed so as to see, as clearly as possible, when its skirmishers in front are attacked; and if it cannot judge by the eye, it will, at least, be able to form an idea of what is passing by the quickness of the fire, or by the hasty retreat of those who oppose the enemy in that quarter where he endeavours to penetrate.

As the march of the enemy must have been foreseen, the commander of a reserve must also have examined and determined how he is to oppose him. He must instantly march, attack with spirit, and dispute obstinately every house, hedge, tree, and ditch, in order to make
known

known the presence of the enemy, and to give time to send support, if circumstances should require it ; or, in other circumstances, that the principal post, and neighbouring ones, may have time to make their dispositions for retreat : for in this case, if one point be carried by the enemy, the others must give way.

When the chain of out-posts is attacked upon all points, and driven back, the light infantry will retire upon such places as have been previously allotted for each detachment, and made known to its commander. These places are the skirts of woods, and the abbatis made for their defence, in front or on the flanks of the field of battle ; hedges, ditches, and houses, under the protection of the line and of the batteries; the intrenchments made on declivities, to defend those parts of the ground which cannot be seen from the batteries, nor by the troops which occupy the summit of the heights.

It is the duty of light infantry to hold out as long as possible in such posts,

s   and

and to defend them to the last extremity, in order to give to the general time to reconnoitre, and ascertain the point upon which the enemy intends to make his principal attack, and to be able to oppose him there with all the means he has contrived to reserve.

In the defence of posts situated on the field of battle, it is no longer necessary to pay so much attention to what is passing on the right and left, as it is in out-posts; for being now under the protection of the line, light infantry has nothing to fear for its retreat, and the battle being now engaged, nothing is to be thought of but obstinate resistance.

SEC-

SECTION III.

Of the Duty of Light Infantry during the Battle.

BATTLES are won or lost by a variety of movements. The manœuvres of light infantry, in these cases, must always be regulated by the movements of the line.

In movements to the front, the light infantry is to take care to cover well the flank of the march. If this flank be attacked, the light infantry must keep its ground, protecting itself by hedges, trees, houses, and every thing which will enable it to resist. It must check the enemy at any risk, and not think of its own safety.

If no enemy present himself on the flank, to attack the troops which march, the light infantry will join their efforts to those of the line, in its attack to the front.

If

and to de...
mity, in or...
to recon...
upon whi...
his princi...
oppose hi...
his contr...

In t...
the field o...
sary to pr...
passing o...
out-posts...
tection of...
thing to f...
being no...
thought...

hedges or other cover in time; and if not, they must instantly run from the circumference to the center, to form a round mass.

Infantry, armed with bayonets, and formed into a close mass, can always resist a charge of cavalry, especially a charge *en fourrageur:* for that purpose, the men must support and press against each other, from the center to the circumference, without breaking the *ensemble.* In this situation, they have only to present the bayonet to the horses' nostrils, and steadily wait for the charge. The cavalry will soon retire beyond the reach of musquetry. The infantry will take the advantage of their retreat to proceed to some cover, without stopping to fire, or load again.

The true defence of infantry against cavalry is in the use of the bayonet, and in the ▪▪▪ of a thick and immoveable bod▪ pressed together *en masse.* either support nor push and the force of one horse may be checked by the united power and weight of seven or eight men.

Musquetry cannot be advantageously used against cavalry, unless the light infantry fire under the protection of cover, where it can load in safety ; for it is evident, that those who are loading their musquets cannot, in that instant, defend themselves, nor take the proper position for resisting cavalry, which is to be done by the pressure and force of several men united against one horse.

As from the nature of its duty, light infantry is often exposed to be attacked unawares by cavalry, it ought to be practised to form quickly into a round mass, whenever it is threatened with a charge of this kind. Having repulsed the charge, the commander will cause it to march in close column to the nearest shelter. Forming the square is not a proper manœuvre for light infantry.

As the fortune of battles cannot be equally balanced for a long time, if part of the line be broken, forced, and thrown

into

into confusion, it would be requiring too much from the light infantry to suppose that it alone could restore order, and renew the conflict: dispersed among the inequalities of the ground, troops of this kind are not capable of a decisive effort.

It is in the nature of the ground, that light infantry, under critical circumstances, must look for resources. The commander must observe the storm, examine upon what ground the enemy advances, his own people retreats, and seize upon the first opportunity to rally and resist; for the first events of a battle are not always decisive, and fortune is often pleased to change sides several times within a few hours.

On every occasion, in all dispositions and situations, the commander of light infantry must keep in reserve a certain proportion of his men. With this reserve he will proceed to the spot where he hopes to be able to make a stand, and will

will cause the retreat to be sounded upon himself.

The first quality of light infantry is to discern the proper time to advance, to resist, and to retire. It is not to allow itself to be thrown into confusion.

A battalion which charges another with the bayonet, either throws the enemy into confusion, or falls into confusion itself. The attacks of light infantry are of a different nature; it fights at open order and scattered, but without disorder or confusion. As it never comes to close quarters with the enemy, it can and ought always to preserve the power of executing whatever manœuvres may be commanded by the bugle-horn.

As the whole of a corps of light infantry is never engaged at once, it can more easily retreat or advance, according to circumstances.

Light infantry, acting parallel with the troops of the line, must resist to the utmost. In advancing, it must advance

with

with them, and cover their wings; and
if they be forced to lose ground, it must
check the enemy at every hedge, wood,
and passage, occasioning him as much
loss as possible. Thus, the light infantry
of a defeated army must endeavour to
join together, and gain a woody height,
village, or defile, where it can stop the
pursuit of the enemy, and cover the retreat of the line.

SEC-

SECTION IV.

Of the Duty of Light Infantry after a Battle.

THE hours which follow either victory or defeat, are equally fatiguing for light infantry: the pursuit of the enemy devolves upon it after a victory, and it must persevere in its endeavours to resist even after the defeat of the line.

In a pursuit, the best way to take prisoners, is not by attacking in front those who are posted and have made dispositions for defence, but by pushing on, and endeavouring to gain ground.

Those who retreat fighting, do right to seek the protection of covered places; but they who pursue must avoid them as much as possible.

It is not requisite to act against a routed army with the caution necessary at the beginning of a battle. Order is

not

not easily restored among troops of the line when they are completely broken.

They have not, in this instance, the advantage of light troops; which, as they are never completely broken, can be more easily rallied. Light infantry must endeavour to advance by the openest ground, and the heights which flank the enemy's retreat.

It is to be remarked, that the march of the enemy is, in general, retarded, by his endeavours to save his artillery. By getting the start of him, and turning his flanks, the light infantry may succeed in occupying a wood, a village, or defile, through which his artillery must pass: it must then make a spirited attack, with as great a noise as possible; little resistance will probably be offered. The enemy, seeing himself thus pursued, will not dare to halt and make a stand, for fear of being cut off entirely, and soon overpowered by superior numbers.

On

On this, as on every other occasion, where it is proper to advance, light infantry ought always to do so by turning the enemy's flank. It is the surest way to find him off his guard somewhere.

It is proper to halt as soon as it gets dark. The commanding officer will take post in some cover by the side of a road. In proportion to the number of his men, he will detach fifteen or twenty to the distance of five or six hundred paces on either flank, with orders to keep up an incessant firing, whether they see an enemy or not. The reason is, that the enemy's stragglers, in order to avoid the firing on the sides, will proceed in the direction where all is apparently quiet, and thus fall into the hands of the detachment concealed in the cover.

At certain seasons, in certain countries, and in certain circumstances, there may be motives for pursuing a defeated army to a greater or less distance; but, at day-break, the officer is not to push

on

on any farther, until he receive orders and instructions to that effect.

From what has been said of a pursuit, the precautions necessary to be taken in a retreat may be deduced.

The light infantry ought to be told off into three grand divisions, one in front, to cover the retreat perpendicularly; the other two on the flanks.

The center one is to occupy, successively, as it has been said before, every place capable of being defended, and to check, as much as possible, the rapidity of the enemy's pursuit, by disputing every inch of the ground, and thereby giving time to the artillery, and perhaps some parts of the baggage, to retire safe.

The divisions on the flanks must be very attentive not to allow themselves to be turned by the enemy. The commanding officers must send on parties to occupy the woods, villages, heights, bridges, and passages; which, if the

T   enemy

enemy were suffered to seize, he might cut off part of the army. These parties must have orders to defend themselves with obstinacy.

At night, the commander of the three divisions of light infantry will endeavour to draw nearer the enemy, keeping, however, a certain number of small detachments before them, which are to conceal themselves, and serve as standing night patroles. In such circumstances, the light infantry ought never to retire from the enemy, unless positively forced to do so, and in danger of being attacked in rear, but continue to observe, and incessantly harass the enemy.

During the night, the commanders of the light infantry, from their knowledge of the country, will contrive to prepare an ambuscade in the rear of their skirmishers, in order to fall upon the enemy with advantage, if he should continue his pursuit at day-break, and

thereby

thereby render him more cautious, and perhaps make him relinquish the pursuit.

Such are, nearly, the different duties of light infantry, in the various occurrences of a battle.

THE END.

Printed by Cox, Son, and Baylis, No. 75, Great Queen
Street, Lincoln's-Inn -Fields.